CLIMATE CHANGE AND ITS CAUSES, EFFECTS AND PREDICTION

CLIMATE CHANGE MEASURES AND TRADE CONSIDERATIONS

CLIMATE CHANGE AND ITS CAUSES, EFFECTS AND PREDICTION

Additional books in this series can be found on Nova's website at:

https://www.novapublishers.com/catalog/index.php?cPath=23_29&seriesp Climate%20Change%20and%20its%20Causes%2C%20Effects%20and %20Prediction&sort =2a&page=1

Additional e-books in this series can be found on Nova's website at:

https://www.novapublishers.com/catalog/index.php?cPath=23_29&seriesp e=Climate+Change+and+its+Causes%2C+Effects+and+Prediction

CLIMATE CHANGE AND ITS CAUSES, EFFECTS AND PREDICTION

CLIMATE CHANGE MEASURES AND TRADE CONSIDERATIONS

GEORGE R. FRIED
EDITOR

Nova Science Publishers, Inc.
New York

For permission to use material from this book please contact us:
Telephone 631-231-7269; Fax 631-231-8175
Web Site: http://www.novapublishers.com

LIBRARY OF CONGRESS CATALOGING-IN-PUBLICATION DATA

Climate change measures and trade considerations / editor, George R. Fried.
p. cm.
Includes index.
ISBN 978-1-60876-756-4 (hardcover)
1.Pollution--Economic aspects. 2. Climatic changes--Economic aspects. 3.
Commercial policy. I. Fried, George R.
HC79.P55C55 2009
363.738'746--dc22

2010012186

Published by Nova Science Publishers, Inc. ✦ New York

CONTENTS

PREFACE

Global climate change is one of the most significant long-term policy challenges facing the United States, and policies to mitigate climate change will have important economic, social, and environmental implications. This book explores the current climate change trade measure considerations for U.S. policy makers, such as estimating industry effects, identifying potentially vulnerable industries and the international trade implications.

Chapter 1 - Elevated levels of greenhouse gases in the atmosphere and the resulting effects on the earth's climate could have significant environmental and economic impacts in the United States and internationally. Potential impacts include rising sea levels and a shift in the intensity and frequency of floods and storms. Proposed responses to climate change include adapting to the possible impacts by planning and improving protective infrastructure, and reducing greenhouse gas emissions directly through regulation or the promotion of low-emissions technologies. Because most U.S. emissions stem from the combustion of fossil fuels such as coal, oil, and natural gas, much of this report centers on the effect emissions regulation could have on the economy.

In this context, GAO was asked to elicit the opinions of experts on (1) actions the Congress might consider to address climate change and what is known about the potential benefits, costs, and uncertainties of these actions and (2) the key strengths and limitations of policies or actions to address climate change. GAO worked with the National Academy of Sciences to identify a panel of noted economists with expertise in analyzing the economic impacts of climate change policies and gathered their opinions through iterative, Web-based questionnaires. The findings reported here represent the views of the 18 economists who responded to both questionnaires.

Chapters 2, 3, 4, 5 and 6 feature testimony before the United States Senate.

In: Climate Change Measures and Trade Considerations
Editor: George R. Fried
ISBN: 978-1-60876-756-4

Chapter 1

CLIMATE CHANGE: EXPERT OPINION ON THE ECONOMICS OF POLICY OPTIONS TO ADDRESS CLIMATE CHANGE*

United State Government Accountability Office

WHY GAO OBTAINED EXPERTS' OPINIONS

Elevated levels of greenhouse gases in the atmosphere and the resulting effects on the earth's climate could have significant environmental and economic impacts in the United States and internationally. Potential impacts include rising sea levels and a shift in the intensity and frequency of floods and storms. Proposed responses to climate change include adapting to the possible impacts by planning and improving protective infrastructure, and reducing greenhouse gas emissions directly through regulation or the promotion of low-emissions technologies. Because most U.S. emissions stem from the combustion of fossil fuels such as coal, oil, and natural gas, much of this report centers on the effect emissions regulation could have on the economy.

In this context, GAO was asked to elicit the opinions of experts on (1) actions the Congress might consider to address climate change and what is known about the potential benefits, costs, and uncertainties of these actions and (2) the key strengths and limitations of policies or actions to address climate change. GAO worked with the National Academy of Sciences to identify a panel of noted economists with expertise in analyzing the economic impacts of climate change policies and gathered their opinions through iterative, Web-based questionnaires. The findings reported here represent the views of the 18 economists who responded to both questionnaires.

* This is an edited, reformatted and augmented version of a U. S. Government Accountability Office publication dated May 2008.

WHAT THE EXPERTS SAID

All of the panelists agreed that the Congress should consider using a market- based mechanism to establish a price on greenhouse gas emissions, and 14 of the 18 panelists recommended additional actions as part of a portfolio to address climate change, such as investment in research and development of low-emissions technologies. Experts differed on the initial stringency of the market-based mechanism, with 14 of the 18 panelists recommending an initial price between less than $1 and $20 per ton of emissions. In addition, 14 of 18 panelists were at least moderately certain that the benefits of their recommended portfolio of actions would outweigh the costs. To establish a price on emissions, most of the panelists preferred either a tax on emissions or a hybrid policy that incorporates features of both a tax and a cap-and-trade program. A tax would set a fixed price on every ton of emissions, whereas a cap-and-trade program would limit or cap total emissions and establish a market for trading (buying and selling) permits to emit a specific amount of greenhouse gases. Under the cap-and-trade system, the market would determine the price of emissions. A hybrid system differs from a traditional cap-and-trade system in that the government would cap emissions, but could sell additional emissions permits if the permit price rose above a predetermined level. Panelists also identified general categories of benefits, such as avoided climate change damages, and costs, such as increases in energy prices, associated with their recommended actions. Overall the panel rated estimates of costs as more useful than estimates of benefits for informing congressional decision making, with some panelists citing uncertainties associated with the future impacts of climate change as limitations to estimating benefits. Further, the majority of panelists agreed that the United States should establish a price on greenhouse gas emissions as soon as possible regardless of the extent to which other countries adopt similar policies. At the same time, the majority of panelists said it was at least somewhat important to participate in international negotiations on climate change.

Panelists identified key strengths and limitations of alternative policy approaches that should be of assistance to the Congress in weighing the potential benefits and costs of different policies for addressing climate change. Many panelists said that a cap-and-trade program would be more effective in achieving a desired level of greenhouse gas emissions because, unlike a tax, it would provide certainty that emissions wouldn't exceed a certain level. However, some of the panelists also said that taxes would be more cost-effective than a cap-and-trade program because the price of emissions would be certain and not susceptible to market fluctuations. Eight panelists therefore preferred a hybrid approach that incorporates features of both a tax and a cap-and-trade program. On average, the panelists rated cost effectiveness as the most important criterion for evaluating various policy options. Finally, panelists said an important strength of using a market-based approach is the ability for the government to raise revenue through a tax or the sale of emissions permits and to use that revenue to offset the adverse effects of the policy.

ABREVIATIONS

CCSP	Climate Change Science Program
CCTP	Climate Change Technology Program

DOE	Department of Energy
EPA	Environmental Protection Agency
EU ETS	European Union Emissions Trading Scheme
IPCC	Intergovernmental Panel on Climate Change
NAS	National Academy of Sciences
NASA	National Aeronautics and Space Administration
NOAA	National Oceanic and Atmospheric Agency
UNFCCC	United Nations Framework Convention on Climate Change
USDA	U. S. Department of Agriculture

May 9, 2008

The Honorable Barbara Boxer
Chairman
Committee on Environment and Public Works
United States Senate

The Honorable Dianne Feinstein
United States Senate

Changes in the earth's climate attributable to increased concentrations of greenhouse gases may have significant environmental and economic impacts in the United States and internationally.[1] Among other potential impacts, climate change could threaten coastal areas with rising sea levels, alter agricultural productivity, and increase the intensity and frequency of floods and tropical storms. Furthermore, climate change has implications for the fiscal health of the federal government, affecting federal crop and flood insurance programs, and placing new stresses on infrastructure and natural resources.[2]

The earth's climate system is driven by energy from the sun and is maintained by complex interactions among the atmosphere, the oceans, and the reflectivity of the earth's surface, among other factors. Certain gases in the earth's atmosphere—such as carbon dioxide and methane— are known as greenhouse gases because they trap energy from the sun and prevent it from returning to space. Climate change is a long-term and global issue because greenhouse gases disperse widely in the atmosphere once emitted and can remain for extended periods of time. According to the Intergovernmental Panel on Climate Change (IPCC)—an organization within the United Nations that assesses scientific, technical, and economic information on the effects of climate change—atmospheric concentrations of carbon dioxide, the most common greenhouse gas, rose 35 percent between pre-industrial times and 2005.[3] The IPCC has determined that 11 of the 12 warmest years on record occurred between 1995 and 2006 and expects that global mean temperatures will continue to rise over the next century as a result of increased atmospheric concentrations of greenhouse gases.

In 2006, carbon dioxide released from the burning of fossil fuels accounted for approximately 78 percent of anthropogenic greenhouse gas emissions in the United States. The remaining 22 percent of emissions included carbon dioxide from nonenergy use of fossil fuels and iron and steel production; methane from landfills, coal mines, oil and gas operations, and agriculture; nitrous oxide from fossil fuels, fertilizers, and industrial processes; and other gases emitted from processes such as refrigeration, air conditioning, and semiconductor manufacturing.[4] In 2005, the United States was the largest global emitter of carbon dioxide followed by China, Russia, Japan, and India.[5]

The United States, 190 other nations, and the European Economic Community have ratified the United Nations Framework Convention on Climate Change (the Framework Convention), which aims to stabilize atmospheric greenhouse gas concentrations within a time frame sufficient to allow ecosystems to adapt naturally to climate change, to ensure that food production is not threatened, and to enable economic development to proceed in a sustainable manner. Under the Kyoto Protocol of the Framework Convention, 178 nations have agreed to reduce their greenhouse gas emissions by at least 5 percent below 1990 levels, by 2012. Also, in 2005, the European Union began implementing its Emissions Trading Scheme (EU ETS), a program that limits emissions in each member state and is intended to help states achieve their commitments under the Kyoto Protocol. Many nations with significant greenhouse gas emissions, including the United States, China, and India, had not committed to binding limits on emissions through the Kyoto Protocol or other mechanisms as of the date of this report. However, in December 2007, the Conference of Parties, the supreme body of the Framework Convention, announced the launch of the Bali Action Plan, a comprehensive process that is expected to lead to a decision in 2009 on steps for countries to take on a post-2012 framework.

Instead of adopting limits on emissions, the United States government has addressed climate change with policies that fall into three main categories: (1) programs targeted at enhancing the scientific understanding of climate change, including the Climate Change Science Program (CCSP), directed by the Assistant Secretary of Commerce for Oceans and Atmosphere; (2) programs that support research, development, and deployment of new technologies that could reduce emissions and improve energy efficiency, including the Climate Change Technology Program (CCTP), led by the Department of Energy; and (3) voluntary programs designed to encourage private and public sector entities to curb their greenhouse gas emissions by providing technical assistance, education, and information sharing, including the Environmental Protection Agency's Climate Leaders Program.

The Congress is currently considering various proposals to further address climate change, including actions to mitigate emissions.[6] Mitigation aims to limit the extent of climate change, usually by decreasing greenhouse gas emissions. One possible mitigation policy is an emissions trading program (referred to as a cap-and-trade program). In general, under a cap-and-trade program, such as the European Union's program, the government would limit the overall amount of greenhouse gas emissions from regulated entities. These entities would need to hold allowances for their emissions, and depending on where the regulation was enforced in the economy, each allowance would entitle them to emit 1 ton of carbon dioxide or to have 1 ton of carbon in the fuel they sold. The government could sell the allowances or give them away (or some combination of the two), and establish a market in which the regulated entities could trade the allowances.[7] For example, firms that find ways to reduce their carbon dioxide emissions below their allowed limit could earn revenue by selling their excess allowances to firms that emit more than their limits.[8] In this manner the market would establish a price for a ton of carbon dioxide emissions based on the supply and demand embodied in such trades. Although the program would provide greater certainty that the level of annual emissions would not increase beyond the emissions cap, the cost of the program could vary, depending on factors such as changes in energy prices. Currently, the United States uses a cap-and-trade program to limit pollutants that cause acid rain emitted by electric utilities.

Another possible mitigation policy is a tax on greenhouse gas emissions. A tax would establish a price on emissions by levying a charge on every ton of carbon dioxide emitted, creating an economic incentive for emitters of greenhouse gases to decrease their emissions by, for example, using fossil fuels more efficiently. Unlike a cap-and-trade program, a tax would provide certainty as to the cost of emitting greenhouse gas emissions, but the precise effect of the tax in reducing emissions would depend on the extent to which producers and consumers respond to higher prices.[9]

The Congress is also considering policies that, unlike a cap-and-trade system or a tax, are not based on establishing a market for greenhouse gas emissions. These options include regulatory approaches, such as standards to increase energy efficiency or the use of renewable energy, and nonregulatory approaches, such as investment in research and development of technologies to reduce emissions. The Congress is also considering measures to adapt to climate change, such as developing protective coastal infrastructure to reduce the impact of rising sea levels. The potential benefits and costs associated with a policy, or combination of policies, depend on factors such as their stringency, timing, and effectiveness of stabilizing or reducing greenhouse gas concentrations.

Much of the debate over the direction of U.S. policy to address climate change has centered on the effects that further policy actions could have on economic growth. For decades, economists have sought to inform this debate by analyzing the potential benefits and costs of actions to address climate change. The benefits of such actions could include avoided damages that may result from changing temperatures. For example, many scientists believe slowing the increase in global mean temperatures and the related rise in sea level may limit damage to coastal areas, which are home to the majority of the U.S. population and account for nearly one- third of the gross domestic product.[10] On the other hand, actions to address climate change would impose costs because most emissions stem from the combustion of fossil fuels, which constitute the majority of the nation's energy supply. Thus, actions to mitigate emissions would likely impose higher costs on producers and users of fossil fuels.

To analyze the economic impacts of different policies for addressing climate change, economists have developed sophisticated models that incorporate historical data on the economic effects of changes in energy prices and assumptions about future economic and climatic conditions. These models focus primarily on the benefits and costs of using market-based mechanisms to impose a price on greenhouse gas emissions and generally place a greater emphasis on analyzing the effect on market goods and services, such as fossil fuels, that have readily available prices, than on analyzing the effect on nonmarket goods such as ecological impacts. Key assumptions underlying these models include the degree of international cooperation in mitigating emissions, the rate of technological change, the sensitivity of the climate to changes in emissions, and the degree to which societies adapt to the impacts of climate change.

In using the models to estimate the benefits and costs associated with controlling greenhouse gas emissions, economists have estimated the potential economic effect of establishing a price on greenhouse gas emissions. According to economic theory, the appropriate emissions price should reflect the social costs that result from emissions.[11] For example, in a survey of the economic literature, the IPCC reported that estimates of the damages associated with current greenhouse gas emissions—impacts on public health, ecosystems, and industry—average about $12 per metric ton of carbon dioxide, with a range from $3 to $95 per ton (2005 dollars).[12] The wide range of estimates primarily reflects differences in the models used and key assumptions.

In this context, you asked us to elicit the opinions of experts in the field of climate change economics on (1) actions the Congress might consider to address climate change and what is known about the potential benefits, costs, and related uncertainties of these actions and (2) the key strengths and limitations of policies or actions to address climate change. To respond to these objectives, we collaborated with the National Academy of Sciences (NAS) to identify and recruit experts with experience analyzing the economic effects of climate change policies. NAS recruited 25 experts affiliated with U.S.-based institutions who have conducted research on the benefits, costs, or uncertainties associated with actions to address climate change, and with in-depth experience in assessing the economic impacts and trade-offs of climate change policies. The experts who served on the panel represent the range of existing research on the economics of climate change, with expertise in areas such as environmental, natural resource, and agricultural economics, and some have served as advisors to the United States government, including as members of the Council of Economic Advisors under the current and former administrations.

To address the first and second objectives, we (1) reviewed relevant climate change academic literature and documents developed by federal agencies and (2) met with agency officials from the Environmental Protection Agency (EPA); the Department of Energy (DOE), including the Energy Information Administration (EIA); the Department of Commerce, including the National Oceanic and Atmospheric Administration (NOAA); the United States Department of Agriculture (USDA); and the National Aeronautics and Space Administration (NASA). To structure our questions and gather opinions from the panelists on our objectives, we used a modified Delphi method, an iterative and controlled feedback approach. With two Web-based questionnaires, we first gathered opinions from the panel on the key topics, and used their responses to develop the second round of questions. In the second round, panelists reacted to the issues and topics discussed in the first round, answering primarily closed-ended questions. We used this approach to eliminate the potential bias associated with live group discussions, and to incorporate more panelists than a live panel would allow. Including more panelists also enabled us to obtain the broadest possible range of opinion. Of the 25 panelists NAS recruited to participate, 21 agreed and were sent the first questionnaire. Nineteen responded to the first questionnaire, and 18 responded to the second. After the responses from the second round were compiled, the panelists were given 2 weeks to comment on a summary of the results. In addition, we followed up with several panelists to verify their responses and elaborate on certain topics. The information presented in this report is primarily from the second questionnaire and represents the views of the 18 experts who participated in both rounds and not GAO's (See app. I for a more detailed description of our scope and methodology). We conducted our work from September 2006 to May 2008.

RESULTS IN BRIEF

All of the economists on the panel agreed that the Congress should consider establishing a price on greenhouse gas emissions using a market- based mechanism but expressed differing views on the type of mechanism and its stringency. In addition, a majority of panelists recommended implementing a portfolio of actions, including at least one complementary policy action in areas such as research and development, adaptation, or international negotiations and assistance. In terms of the mechanism for establishing a price, 8 panelists preferred a cap-and-trade system with the government having the ability to use a cost control mechanism (called a safety valve) if the price of permits exceeds certain levels, 7 preferred a tax on emissions, and 3 preferred a cap-and-trade system without the safety valve. Despite key uncertainties associated with estimating potential costs and benefits, 14 of the 18 panelists said they were at least moderately certain that the benefits of their preferred portfolio of actions would outweigh the costs, and 4 did not respond to questions on this topic. With respect to the stringency of their recommended market-based mechanism, 7 panelists said the price per ton of emissions should range from less than $1 to $10, 7 said from $11 to $20, and 3 said it should exceed $20 (2007 dollars).[13] In addition, the majority of panelists said that the price on emissions should gradually increase over time. Further, all of the panelists said that the price should be implemented by 2015, and that it should apply to all sectors of the economy.

Panelists identified general categories of potential benefits and costs and provided some quantitative estimates or cited estimates from related studies. The panelists rated avoided climate change damages as the most important type of benefits; these damages may include flooding from rising sea levels and extreme weather events. Some panelists also provided cost estimates for their preferred actions and cited key studies underlying those estimates. Overall, the panelists said that estimates of costs are more useful for informing congressional decision makers than estimates of benefits, but all of the panelists that responded to the applicable questions on this topic said that the estimates of costs and benefits from integrated assessment models were at least somewhat useful. The panelists identified key uncertainties that affect these estimates, including uncertainty about the extent to which rising temperatures could lead to abrupt changes in the climate system, the science of climate change, and the potential economic effects of actions to address climate change. Despite these uncertainties, 16 of the 18 panelists agreed that the United States should limit emissions as soon as possible, regardless of the efforts of other nations to adopt similar policies. At the same time, the majority of the panelists said that it was at least somewhat important to participate in international negotiations, either to facilitate climate agreements or to enhance the credibility or influence of the United States.

Panelists identified key strengths and weaknesses of alternative policy approaches that should be of assistance to the Congress in weighing the potential benefits and costs of different policies for addressing climate change. Notably, the experts discussed the greater certainty of attaining emissions targets under a cap-and-trade system versus the greater efficiency of a tax in achieving emissions reductions at a lower cost. On average, they rated cost-effectiveness as the most important criterion for evaluating various policy options, and used it and other important criteria to compare the strengths and limitations of different actions to address climate change. Some panelists said that a cap-and-trade program would be more effective in achieving a desired level of greenhouse gas emissions because, unlike a tax, it would provide certainty that emissions would not exceed a certain level. However, some of the panelists also said that taxes were more economically efficient than a cap-and-trade program because the price of emissions would be certain and would not be susceptible to market fluctuations that could lead to increased costs. In addition, some panelists felt a cap could be more administratively burdensome than a tax.

Nonetheless, some of the panelists that preferred a tax said that a cap-andtrade program, especially if it included cost-minimizing components, would be an acceptable second option to address climate change. Eight panelists preferred a hybrid approach where the government would create a cap-and-trade system with the option of selling additional permits if the market price of permits exceeded a certain level, thereby minimizing risks of adverse economic consequences. However, the emissions reductions achieved by a hybrid program would be less certain than a standard capand-trade program if the price control mechanisms came into effect. As for distributing emissions permits under a cap-and-trade or hybrid program, the majority of panelists favored at least partial auctioning of permits rather than free distribution. They noted that the government could redistribute revenue from permit auctions to offset adverse effects on consumers or particular sectors of the economy. The panelists also discussed the strengths and limitations of other policy options, including research and development of technologies, adaptation to the anticipated impacts of climate change, revising energy efficiency standards, and reforming subsidies for fossil fuel production and other industries.

BACKGROUND

Greenhouse gases—including carbon dioxide, methane, nitrous oxide, and other substances—trap a portion of the sun's heat in the atmosphere and prevent the heat from returning to space. The insulating effect, known as the greenhouse effect, moderates atmospheric temperatures, keeping the earth warm enough to support life. According to the Intergovernmental Panel on Climate Change (IPCC), global atmospheric concentrations of these greenhouse gases have increased markedly as a result of human activities over the past 200 years, contributing to a warming of the earth's climate.[14]

The IPCC generally attributes increases in average global air and ocean temperatures, widespread melting of snow and ice, and rising mean global sea levels to a warming of the earth's climate system. Furthermore, according to the IPCC, the oceans have absorbed more than 80 percent of the heat added to the earth's climate system, causing seawater to expand, thereby contributing to sea level rise. Scientists have also reported that mountain glaciers and snow cover have declined, on average, in both hemispheres, and that widespread decreases in the sizes of glaciers and polar ice caps, combined with losses in the ice sheets of Greenland and Antarctica very likely contributed to a sea level rise of 0.17 meters during the 20th century.

Table 1. Potential Impacts of Climate Change by Sector

Sector	Major projected impacts
Agriculture, forestry, and ecosystems	• Increased yields in colder environments • Decreased yields in warmer environments • Increased insect outbreaks • Increased danger of wildfires • Damage to crops • Waterlogging of soils • Land degradation • Increased livestock deaths • Uprooting of trees • Damage to coral reefs • Salinization of irrigation water, estuaries, and freshwater systems
Water resources	• Effects on some water resources, such as increased salinization of groundwater anddecreased availability of freshwater for humans and ecosystems • Increased water demand • Water quality problems • Adverse effects on quality of surface and groundwater • More widespread water scarcity • Power outages causing disruption of public water supply • Decreased freshwater availability due to saltwater intrusion
Human health	• Reduced human mortality from decreased cold exposure • Increased risk of heat-related mortality • Increased risk of deaths, injuries, and infectious respiratory and skin diseases • Increased risk of food and water shortage • Increased risk of malnutrition • Increased risk of water- and food-borne diseases • Increased risk of deaths and injuries by drowning and floods
Industry, settlement, and society	• Reduced energy demand for heating • Increased energy demand for cooling • Declining air quality in cities • Reduced disruption to transport due to snow, ice • Disruption of settlements, commerce, transport, and societies due to flooding • Pressures on urban and rural infrastructures • Water shortages for settlements, industry, and societies • Reduced hydropower generation potential • Potential for population migration • Disruption by flood and high winds • Withdrawal of risk coverage in vulnerable areas by private insurers • Costs of coastal protection versus costs of land use relocation • Potential for movement of populations and infrastructure

Source: IPCC, Working Group III, AR4, Summary for Policymakers

The effect of increases in atmospheric concentrations of greenhouse gases and temperature on ecosystems and economic growth is expected to vary across regions, countries, and economic sectors (see table 1).

For example, small island nations are particularly at risk because of their vulnerability to sea level rise, poor coastal infrastructure, and economies that rely heavily on coastal fishing and tourism. Alternatively, while certain areas of the United States may be adversely affected by rising sea levels, its diverse economy, significant resources, and established infrastructure may help moderate the negative effects associated with climate change. Figure 1 shows a selection of projected impacts from climate change on different regions, assuming that greenhouse gas emissions and concentrations continue to increase at current rates.

According to the IPCC, in 2004, developed countries, including the United States, constituted 20 percent of global population, but were responsible for nearly half of global greenhouse gas emissions.[15] However, in the absence of mitigation policies, the IPCC projects that between 2000 and 2030, two-thirds to three-quarters of the projected increase in global carbon dioxide emissions will occur in developing countries.[16]

Mitigating Greenhouse Gas Emissions

Many developed countries have begun to mitigate or reduce greenhouse gas emissions by adopting policies such as carbon taxes, cap-and-trade programs, energy efficiency standards, financial incentives (e.g., subsidies or tax credits), voluntary agreements, education campaigns, and research, development, and deployment of advanced technologies.[17] For example, the European Union Emission Trading Scheme (EU ETS) is a cap-andtrade system in which energy-intensive industries in the European Union buy or sell emission allowances to help meet member states' commitments under the Kyoto Protocol. The EU ETS covers over 11,000 energy- intensive installations, such as oil refineries and steel plants, in 25 member countries and covering nearly half of Europe's carbon dioxide emissions. Governments may also use a portfolio of policies; for example, a cap-andtrade system may be pursued in combination with energy efficiency standards and financial incentives for certain sectors. Table 2 shows selected policies and instruments that have been shown to be environmentally effective at a national level.

In addition, carbon capture and storage can supplement other climate change mitigation policies. Carbon capture and storage involves separating and storing carbon dioxide from an industrial or energy-related source, thereby preventing carbon dioxide emissions into the atmosphere. Carbon capture and storage is most commonly used to enhance oil and gas recovery in depleted fields. Efforts to capture carbon dioxide from power generation or industrial processes are currently the main focus of research and development of the technology. This process has the potential to reduce emissions, but its widespread use may be limited by several barriers, including technological feasibility, costs, regulatory issues, and environmental concerns.

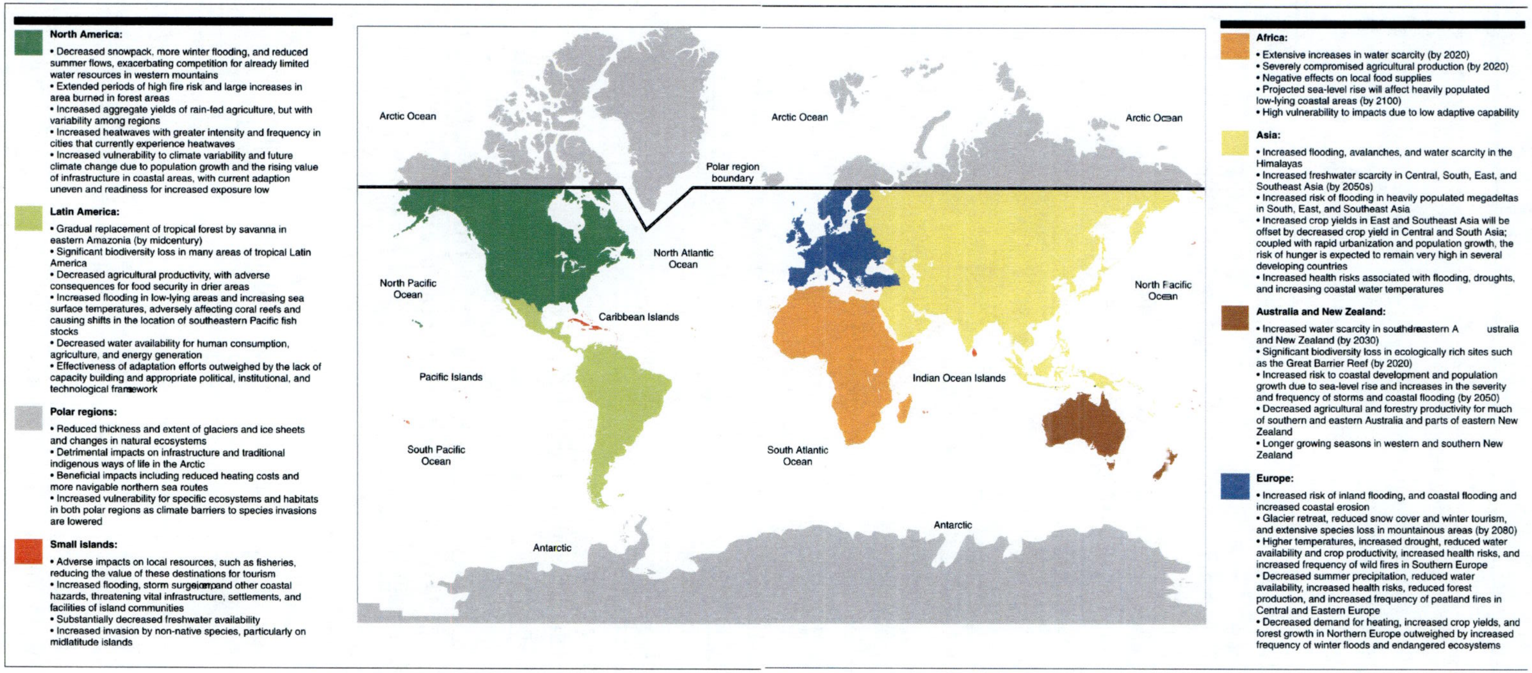

Figure 1. Potential Impacts of Climate Change by Geographic Region.

Table 2. Selected Policies, Measures, and Instruments Currently Used by Various Nations to Address Climate Change

Sector	Policies, measures, and instruments
Energy Supply	• Reduction of fossil fuel subsidies • Taxes or carbon charges on fossil fuels • Incentives for renewable energy • Public research, development, and deployment of low-emission technologies
Transportation	• Mandatory fuel economy standards, biofuel blending, and CO_2 standards for road transport • Taxes on vehicle purchase, registration, and use, and on motor fuels, road and parking pricing • Land use regulations and infrastructure planning • Investment in public transport facilities and nonmotorized forms of transport • Public research, development, and deployment investment in low-emission technologies
Buildings	• Appliance standards and labeling • Building codes and certification • Demand-side management programs to incentivize customers to purchase energy-efficient products • Public sector leadership programs, including procurement requirements for governments • Incentives for energy service companies • Public research, development, and deployment investment in low-emission technologies
Industry	• Performance standards • Subsidies, tax credits • Tradable permits • Voluntary agreements • Public research, development, and deployment investment in low-emission technologies
Agriculture	• Financial incentives and regulations for improved land management, maintaining soil carbon content, efficient use of fertilizers and irrigation • Public research, development, and deployment investment in low-emission technologies
Forestry	• Financial incentives (national and international) to increase forest area, to reduce deforestation, and to maintain and manage forests • Land use regulation and enforcement • Public research, development, and deployment investment in low-emission technologies
Waste management	• Financial incentives for improved waste and wastewater management • Renewable energy incentives or obligations • Waste management regulations • Public research, development, and deployment investment in low emission technologies

Source: IPCC, Working Group III, AR4, Summary for Policymakers.

Another option is to allow for carbon offsets, which is a way for consumers and producers to compensate for greenhouse gas emissions occurring in one location by reducing or avoiding emissions somewhere else. For example, a manufacturing facility in the United States could compensate for its emissions by purchasing carbon offsets from a tree- planting project in South America. Carbon offsets are traded in compliance markets, to satisfy requirements to limit emissions, and in voluntary markets, where emissions reductions are not required but may serve other purposes. For example, carbon offsets serve as a mechanism for complying with the emissions reduction requirements of the EU ETS. Under this scheme, certain regulated entities may choose to comply with emissions limits by purchasing offsets rather than by reducing their own emissions.

Adaptation Policies to Reduce Vulnerability to Climate Change

In addition to mitigating greenhouse gas emissions, policies to adapt to climate change could help reduce the vulnerability of countries and regions to potentially adverse impacts. For example, raising river or coastal dikes could protect coastal communities and resources from sea level rise. The vulnerability of a country or region depends on both on the susceptibility of a political, economic, or natural system to the adverse effects of climate change and the capacity of a society to adjust to the expected change. For example, less developed economies may face difficulty adapting to climate change because of poor infrastructure, poverty, and resource constraints. Adaptation may be viewed as a risk-management strategy for protecting vulnerable countries, sectors, and communities that might be affected by changes in the climate and related impacts.

In December 2007, members of the Conference of the Parties to the Framework Convention agreed to launch a comprehensive process that will lead to the adoption of a decision in 2009 on next steps for countries to take on climate change. This process, called the Bali Action Plan, includes provisions that require signatories, including the United States, to undertake efforts to enhance international cooperation on adaptation, such as vulnerability assessments; capacity building; risk management; and reduction strategies; and the integration of adaptation into planning decisions.[18]

Estimating the Potential Benefits and Costs of Actions to Address Climate Change

Economists and other researchers have developed integrated assessment models that utilize economic and climate and other environmental data to assess the economic consequences associated with different policies for addressing climate change. These models vary in structure and scope, but generally include historical data on the U.S. and international economies, emissions and atmospheric concentrations of greenhouse gases, and global temperature. Typically, economists use the models to estimate the economic consequences of policy actions, by comparing the present values of the economic costs of an action such as tax or cap and trade and the future benefits it would be expected to generate, relative to a businessas-usual emissions projection (for example, no significant reductions in greenhouse gas emissions). In judging whether a policy action would be preferable, economists evaluate policies using criteria such as (1) economic efficiency, where the action maximizes potential net benefits (total benefits minus total costs), compared to business as usual and (2) cost-effectiveness, where the action achieves the chosen policy objective, such as an emissions reduction target, at least cost.

In general, the economic costs that society would incur from taking action to address climate change will begin to occur immediately, while the economic and environmental benefits will mainly occur decades in the future as atmospheric concentrations of greenhouse gases and global temperatures stabilize. The economic costs of taking action represent the value of the goods and services that society would forgo to allocate resources to the emissions control policy, including compliance expenditures, administration and enforcement costs, and other costs that the action might impose on the economy (for example, as a result of higher prices). For example, because the energy sector and energy- intensive industries generate substantial emissions of greenhouse gases, a significant component of the potential cost associated with actions to address climate change relates to the impacts of changes in energy prices. The economic costs of reducing emissions to stabilize atmospheric greenhouse gas concentrations will increase as the stringency of emissions reduction goals increase, and correspondingly, as the stabilization goal decreases.

The potential economic benefits of policies to address climate change generally consist of the effects of stabilizing or reducing the atmospheric concentration of greenhouse gases and global temperature on human welfare. Typically, benefits are measured in terms of the damages that would be averted if an action were taken to address climate change. For example, under a business-as-usual scenario, researchers project that further increases in atmospheric concentrations of greenhouse gases and global temperature could reduce agricultural productivity in certain parts of the world, increase the incidence of diseases in certain climates, and reduce the environmental goods and services provided by some ecosystems. In addition, researchers have estimated that limiting greenhouse gas emissions and slowing the increase in atmospheric concentrations and global temperature would avoid some damages.

Estimating the potential benefits associated with actions to address climate change, however, can pose challenges, partly because of uncertainty about the magnitude of climate impacts and the resulting effect on human welfare. In particular, scientists face challenges estimating the effects of climate change at the regional and local level. In addition, rising global temperature could involve unexpectedly abrupt changes in the climate, which would be more costly than if changes are more moderate. According to NAS, global warming and other human alterations of the earth's climate system may increase the possibility of large and abrupt regional or global climatic events. However, NAS concluded that because the abrupt climate changes of the past are not yet fully explained, future abrupt changes cannot be predicted with any confidence and climate surprises are to be expected.[19] In addition, partly because many of the environmental goods and services expected to be affected by climate change are generally not bought and sold in markets, it is difficult to develop reliable estimates of the value associated with the damages expected from climate change.[20]

Furthermore, estimating the economic consequences of various actions to address climate change involves consideration of the fact that the benefits and costs will occur in different time periods and should be expressed in comparable present value terms. For example, because greenhouse gases accumulate in the atmosphere over very long periods of time, the effect of current actions to stabilize concentrations and global temperature might take decades to manifest themselves. As a result, the future benefits and costs of actions to address climate change are typically discounted (by using a discount or interest rate) to estimate their present value. In general, discounting is used to reflect the extent to which individuals trade off current for future consumption. Nonetheless, because discounting generally attaches a lower weight to future impacts compared to near-term impacts, the choice of the discount rate is an important factor in assessing the economic consequences of actions to address climate change.[21]

Moreover, although the cost of any U.S. action to address climate change will be primarily borne by U.S. producers and consumers, the economic benefits associated with the action would also accrue to other countries that face greater vulnerability to climate change. According to the IPCC, the global distribution of climate change impacts varies throughout the world and nonclimate stresses such as poverty and food insecurity can increase a country's vulnerability to climate change by reducing its resilience and capacity to adapt. For example, the IPCC estimated that Africa could face increased risk of water scarcity and reduced food security.[22] However, because African nations account for a negligible share of total greenhouse gas emissions, it is expected that they will bear a much smaller portion of the overall costs or responsibility for greenhouse gas mitigation.

Climate Change Policies Currently under Consideration by the Congress

The eight climate change mitigation bills currently under consideration by the Congress provide an overview of the potential for simultaneously pursuing a portfolio of actions. All eight bills include provisions for a capand-trade system in combination with initiatives to promote the development and adoption of low-carbon technologies.[23] Also, three bills would require that a specific amount of electricity be generated by renewable energy—generally called a Renewable Portfolio Standard— including wind and solar energy, and energy efficiency performance standards. Five bills would apply the emissions caps to specific sectors of the economy, such as electricity, transportation, and industry, while caps under the other three would not be limited to specific economic sectors.

DESPITE SOME UNCERTAINTY REGARDING THE POTENTIAL ECONOMIC IMPACT, ALL OF THE PANELISTS SUPPORTED ESTABLISHING A PRICE ON GREENHOUSE GAS EMISSIONS

All of the panelists agreed that the Congress should consider establishing a price on greenhouse gas emissions using a market-based mechanism, but they expressed differing views on the type of mechanism and its stringency. In addition, 14 of the 18 panelists were at least moderately certain that the benefits of their suggested portfolio of actions would outweigh the costs.[24] Most of the panelists identified either a tax on emissions or a cap-and-trade program with a safety valve as the preferred mechanism to establish a price on emissions, and the majority believed a portfolio of additional actions to address climate change could complement the market-based mechanism. Some panelists also identified general categories of benefits and costs associated with their recommended actions, and rated the usefulness of benefit and cost estimates derived from integrated assessment models. Overall the panel rated estimates of costs as more useful than estimates of benefits for informing congressional decision making, with some panelists citing uncertainties associated with the future impacts of climate change as a limitation to estimating benefits. Finally, while some panelists said that the United States should proceed cautiously if it acts unilaterally, the majority of panelists agreed that it should establish a price on greenhouse gas emissions as soon as possible regardless of the extent to which other countries adopt similar policies.

All of the Panelists Agreed That the Congress Should Consider Establishing a Price on Greenhouse Gas Emissions, and the Majority Recommended Complementary Policies

All of the panelists agreed that the Congress should consider a market- based mechanism to establish a price on greenhouse gas emissions and supported implementation of the policy by 2015. Opinions varied on whether the Congress should implement a cap-and-trade system or a tax to control greenhouse gas emissions, with eight panelists preferring a capand-trade program with a safety valve (sometimes referred to as a hybrid system), seven preferring a tax, and three preferring a cap-and-trade program. All of the panelists agreed that the policy should target all sectors of the economy, and the majority believed that it should include all greenhouse gases. For example, one panelist stated that by establishing a price on emissions from all sources in the United States with no exceptions, the policy would equilibrate the marginal cost of reducing emissions across all sources, making it economically efficient.

The panelists varied in their views on the stringency of the market-based regulatory mechanism that they supported to place a price on greenhouse gas emissions.[25] For example, in proposing an initial price on emissions, seven panelists said it should range from less than $1 to $10, six said from $11 to $20, and four said it should be greater than $20 (2007 dollars per metric ton of carbon dioxide equivalent).[26] In addition, while most panelists said the price should increase over time, they varied in their views on the preferred rate of increase. For example, some panelists provided estimates ranging from 2 percent to 5 percent per year adjusted for inflation, while another panelist said more generally that it should be reevaluated periodically (for example, every 5 years) and rise as marginal damages of climate change rise. Some panelists noted the importance of a long-term commitment to establish a price on emissions and the flexibility to adjust the price and rate of increase as new information becomes available. For example, one panelist stated that certainty in setting emissions reductions goals was necessary for firms that would have to make substantial investments in new emissions reduction technologies.

Another panelist added that while the program should run into the distant future, it should include an explicit feature to revise taxes or emissions targets based on near-term mitigation experience, new research on climate change science and impacts, and actions by other countries.

Under a hybrid approach, the government would establish a cap-and-trade system with a safety valve, a mechanism under which the government would sell additional permits if the market price exceeded a predetermined level, which would represent the maximum permit price and thus an upper limit on control costs. A safety valve price would be established, which would represent the maximum permit price and thus an upper limit on control costs. Panelists preferring a hybrid policy varied in their recommendations for an initial safety valve price, with five preferring a safety valve price identical or close to their targeted initial market price for greenhouse gas emissions and two recommending a price moderately or well above their recommended market price (see app. II for additional information on this topic). For example, a panelist recommending a targeted initial market price of $0.55 per metric ton of emissions (in carbon dioxide equivalent) recommended an identical safety valve price, adding that the price should not be much higher than the targeted marginal cost of reducing emissions. Another panelist recommending an initial market price between $11 and $20 said that the safety valve should depend on the stringency of the policy. When providing an example, the panelist added that if the initial market price per metric ton of emissions was $20, then the safety valve should be set three to four times above the initial price of emissions, or between $60 and $80 (See app. II for additional information on this topic).

In addition, seven of the eight panelists that prefer a hybrid policy believe that the price of the safety valve should increase over time, and the majority of panelists said that the safety valve should be reevaluated periodically based on new information. For example, one panelist said that the safety valve price should increase by 2.5 percent plus inflation annually, and that the government should reevaluate the stringency of the policy every 5 to 10 years based on new information. Another panelist said that the safety valve provision could be abandoned altogether after a periodic review of the adequacy of the policy and its costs.

Overall, eight of the panelists said that a market-based mechanism should be imposed "upstream," where fossil fuels first enter the economy, four preferred a "downstream" mechanism that regulated direct and indirect emitters, and five preferred a mechanism with both upstream and downstream components. An upstream system would require fossil fuel producers, such as extractors and processors, and importers, to pay a tax or to hold permits based on the carbon content of the fuels. Alternatively, a downstream system would regulate sources such as electric utilities that combust fossil fuels and emit greenhouse gases. While environmental effectiveness would likely be the same under either approach, one panelist said that an upstream policy would have lower administrative costs when compared to a downstream system partly because it involves a much more manageable set of firms. Of the panelists who said they preferred a combination of upstream and downstream provisions, three preferred a policy that had downstream provisions for carbon capture and storage from utilities or other processes that remove carbon from the atmosphere. See appendix II for additional information on experts' preferred design for a market-based policy.

In the second round questionnaire, we asked panelists to rate the importance of complementary actions to address climate change that had been recommended in the first round. We also asked panelists to identify their recommended portfolio of actions for the Congress to consider. When rating the importance of recommended actions, the panelists gave the highest average ratings to funding of research, development, and deployment of zero-carbon and low-carbon technologies, and participation in international negotiations (see figure 2 and app. III for more detail). For example, recognizing that the private sector may not invest in some technologies, one panelist recommended public investment to accelerate development of carbon capture and storage from electricity generation. This panelist also advocated expanded research into climate change adaptation in the southern United States and in developing countries for agriculture, forestry, and fisheries, emphasizing the importance of preparing for inevitable climate impacts. Another panelist recommended that the United States engage in international negotiations to make similar emissions reductions commitments, and added that the United States should consider helping other countries meet their targets as a part of international aid.

When recommending a portfolio of policy options for the Congress to consider, 14 of the 18 panelists identified actions to address climate change in addition to placing a price on greenhouse gas emissions (see app II, table 4, for more detail). Of the 14 holding this view, 10 said research and development in low- or zero-carbon technologies or research in the basic science of climate change should be part of the portfolio, 7 said international negotiations or assistance to developing countries should be included, and 6 said adaptation should factor into a portfolio of actions (see app. II, table 4, for more detail). When asked how certain they were that their recommended portfolio of actions to address climate change were economically justified, 14 of the 18 panelists were at least moderately certain that the benefits of their suggested actions would outweigh the costs. The remaining 4 panelists did not know or did not provide a response to this question.

Panelists Described the Potential Benefits and Costs of Actions to Address Climate Change

In the first questionnaire, we asked panelists to identify potential categories of costs and benefits associated with actions to address climate change. In the second questionnaire, we asked them to rate the importance of categories of benefits as a rationale for addressing climate change.[27] On average, panelists rated avoiding damages such as those from flooding, impacts on sensitive ecosystems, public health, and species loss as the most important category of potential benefits (see Figure 3 and app. III for more detail). In addition, panelists rated reducing risk of extreme or irreversible climate events as the second most important category of potential benefits. For example, one panelist discussed the benefits of reducing the probability of abrupt or catastrophic climate events such as dramatic sea level rise by stabilizing atmospheric concentrations of greenhouse gases. Another noted the importance of avoiding damages by reducing the risks of vulnerability to water scarcity, hunger, or the frequency of storm events.

The panelists also rated establishing a price signal to influence market or individual behavior and facilitating international cooperation on climate change as an important category of benefits that could serve as a rationale for actions to address climate change. For example, one panelist stated that the United States needs an unavoidable price signal to fully harness the innovativeness of the U.S. industrial and scientific communities and to provide incentives to reduce emissions. Another panelist stated that a modest near-term market-based policy could provide an opportunity to learn about the effectiveness of an emissions tax or cap-and-trade program, and stimulate research to inform the stringency of future policies.

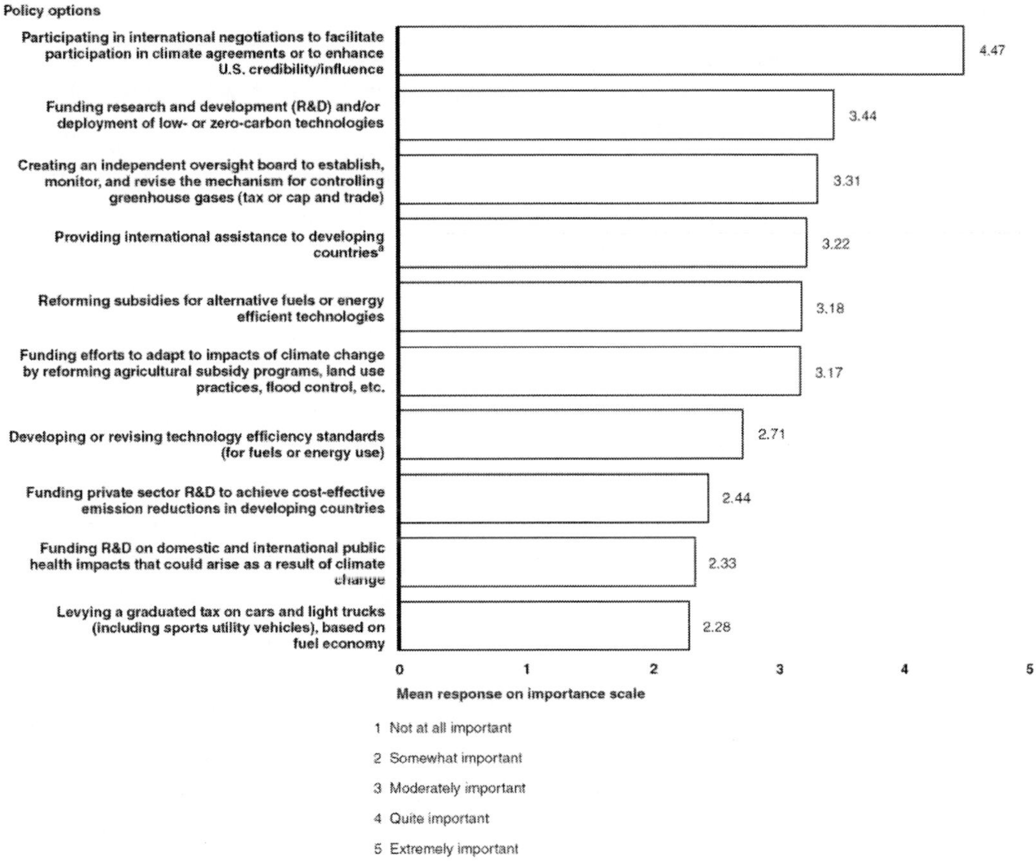

Source: GAO analysis.

Note: Total responses for each question may range from 15 to 18. See appendix III for additional details.

[a] Providing international assistance would include compensation for climate change impacts or assistance deploying low-carbon technologies.

Figure 2. Mean Panelist Ratings of the Importance of Additional Policy Options to Address Climate Change

In the first questionnaire, some panelists presented estimates of the costs that would be associated with their proposed actions to address climate change. Panelists' estimates of the impacts on social welfare, including the effects on economic growth, varied depending on the type and stringency of the policy recommended. Some panelists cited their own research and other academic studies, as well as several assessment reports by domestic and international governmental entities, as credible estimates of the economic costs associated with their proposed actions, but some noted that policy choice and stringency can have a large impact on the cost estimates.

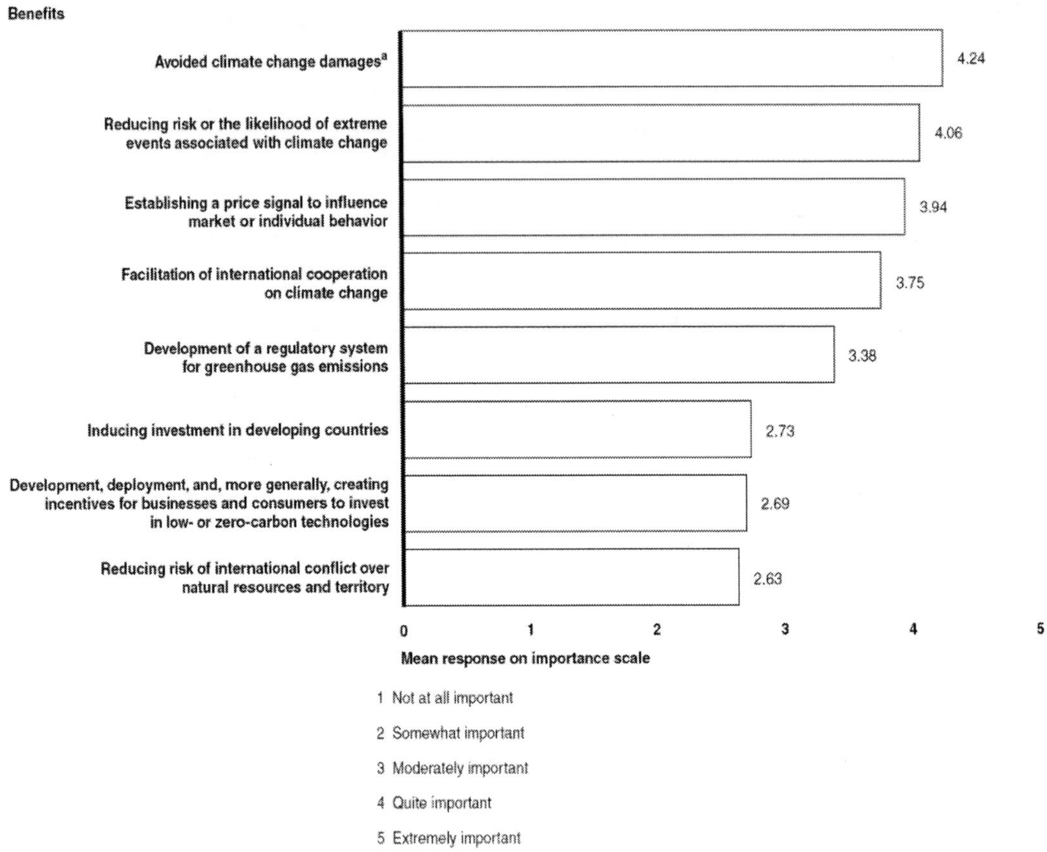

Source: GAO analysis.

Note: Total responses for each question may range from 15 to 18. See appendix III for additional details.

[a] Avoided climate change damages includes flooding, extreme weather events, crop damage, impacts on sensitive ecosystems, public health, and species loss.

Figure 3. Mean Panelist Ratings of the Importance of General Categories of Potential Benefits as a Rationale for Addressing Climate Change

For example, one panelist cited a recent modeling effort by the U.S. Climate Change Science Program (CCSP) that projected that by 2060, gross world product (GWP) would decline by between 0 and 6.7 percent per year under emissions reduction scenarios that achieved different levels of atmospheric stabilization of greenhouse gases.[28] However, another panelist stated that welfare cost estimates from the study could be reduced by as much as 50 percent by incorporating certain policy features such as carbon offsets into the study's policy scenario. Further, another panelist added that it is difficult to provide a comprehensive list of potential costs given the breadth of possible policy scenarios.

Some panelists also commented on the effect of their suggested actions on energy prices. For example, one panelist said that a mandatory emissions reduction program that establishes a price around $20 per ton of carbon dioxide (2005 dollars) would increase gasoline prices by approximately 20 cents per gallon and residential electricity prices by approximately 1 cent per kilowatt hour above business as usual estimates. Another expert estimated that a $20 price per ton of carbon dioxide equivalent would result in a 10 percent increase in consumer energy prices. Both experts cited a recent study by the Energy Information Administration on energy market impacts of alternative greenhouse gas reduction policies to support their estimates.

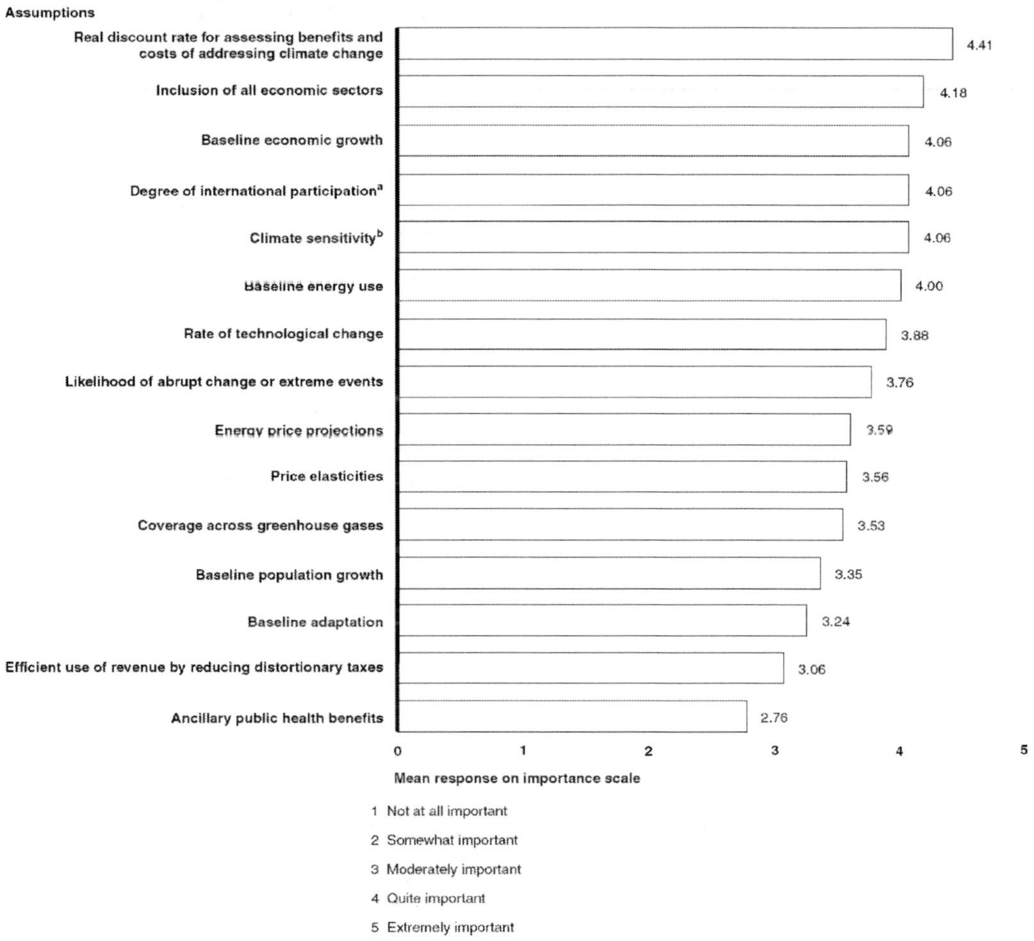

Source: GAO analysis.

Note: Total responses for each question may range from 15 to 18. See appendix III for additional details.

[a] Degree of international participation means how quickly and to what extent high-emitting nations implement an emissions reduction policy.

[b] Climate sensitivity is the change in global mean temperature that results when the climate system attains a new equilibrium as the result of a doubling in the atmospheric carbon dioxide concentration.

Figure 4. Mean Panelist Ratings of the Importance of Types of Assumptions in Integrated Assessment Models

Finally, in responding to our first questionnaire, panelists were asked to identify key assumptions that they made in describing the estimates of the benefits and costs of their proposed actions to address climate change. In the second questionnaire, the panelists rated the importance of the assumptions in terms of affecting the benefits and costs estimates generated by integrated assessment models (see app. III for more detail). Figure 4 illustrates the opinion of panelists on the importance of various assumptions made in developing estimates using integrated assessment models. Assumptions that panelists identified as the most important were the real discount rate (interest rate used for discounting, adjusted for inflation) for assessing the benefits and costs of climate change, and the inclusion of all economic sectors in policies to address climate change. With respect to discounting future benefits and costs, we asked panelists to identify a reasonable estimate for the discount rate. Fifteen panelists responded with estimates ranging from 0 to 5 percent.

Panelists Rated Estimates of Costs of Actions to Address Climate Change as More Useful than Estimates of Benefits, Citing Uncertainties Associated with Future Impacts

Citing uncertainties associated with the potential future impacts of climate change, and the difficulties of estimating their economic impacts, panelists rated cost estimates from integrated assessment models as more useful for informing congressional decision making than benefit estimates. While panelists identified challenges in estimating costs such as predicting future technological development and additional costs associated with inefficient policy designs, 10 panelists stated that the estimates of costs were quite or extremely useful, whereas the majority of panelists said that estimates of benefits were only somewhat or moderately useful. However, all of the panelists that responded to the applicable questions on this topic said that the estimates of costs and benefits from integrated assessment models were at least somewhat useful. Panelists provided a number of rationales for their opinions on integrated assessment models. For example, one panelist said that costs were easier to estimate primarily because economists have historical data that can be used to model the effect of changes in energy prices. Conversely, one panelist stated that the benefits of emissions mitigation are poorly understood because they are based on a limited understanding about the climate system. Another panelist added that even though researchers have put substantial effort into quantifying the avoided damages of actions to address climate change, they are still highly speculative.

When asked to rate the relative importance of uncertainties that may affect the estimated benefits and costs in integrated assessment models identified in the first questionnaire, on average, the panelists rated thresholds and abrupt changes in the climate system as the most important uncertainty, followed by the science of climate change, and the economic effect of actions to address climate change (see Figure 5 and app. III for more detail).[29] Additionally, the panelists identified how society will adapt to climate change as an important uncertainty affecting benefit and cost estimates. When asked how to best address risk and uncertainty in economic assessments of climate change policies, two panelists said it was important to provide additional information on uncertainties associated with low-probability, high-impact climate change events. For example, one panelist said that the limited treatment of low-probability high-impact uncertainty associated with high temperature changes is the biggest unsolved problem driving a complete economic analysis. Other panelists said it was important to provide decision makers with descriptions of the risks of climate change, including distributions of probable impact scenarios under various policy approaches. For example, one panelist said that ideally, economic assessments should fully characterize uncertainties by providing decision makers with probability distributions over prospective outcomes.

Despite the challenges associated with estimating benefits and costs using integrated assessment models, several panelists cited them as valuable tools for informing judgment on climate change policy. For example, one panelist said that integrated assessment models are the only tools that can assemble all of the factors surrounding the climate issue to assess potential emissions mitigation and adaptation strategies. When asked to identify steps that the Congress could take to help economists and other researchers address the most important uncertainties, most of the panelists that responded expressed the need for continued funding of research to improve understanding of the science and economics of climate change.

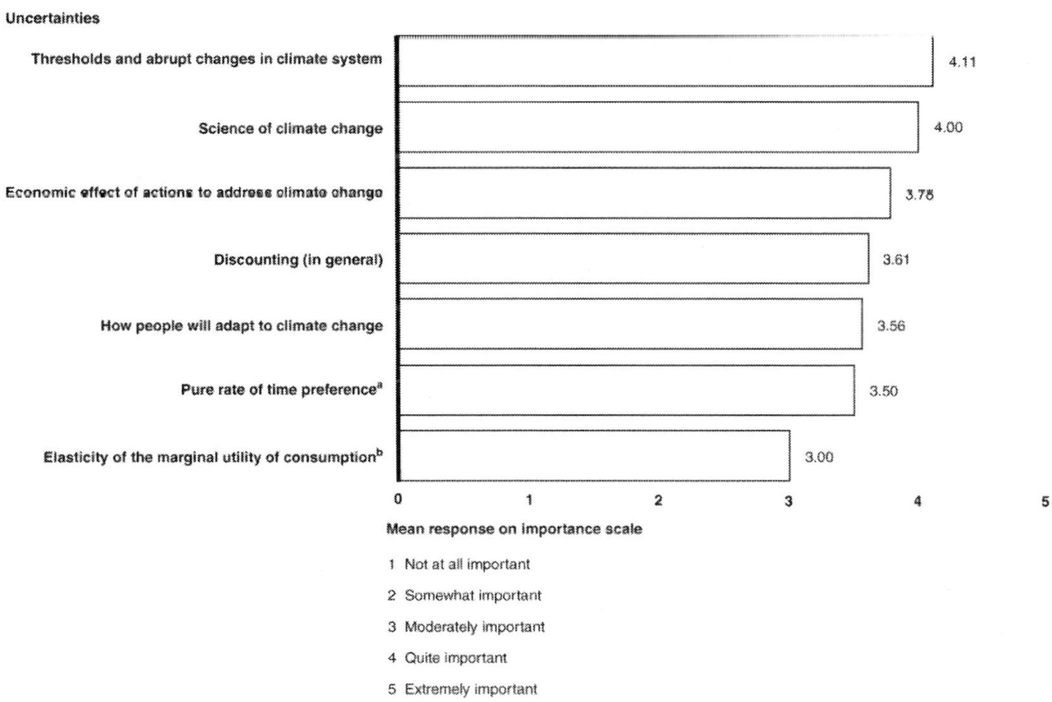

Source: GAO analysis.

Note: Total responses for each question may range from 15 to 18. See appendix III for additional details.

[a] Pure rate of time preference reflects the relative weight assigned to the welfare of different generations over time.

[b] Elasticity of the marginal utility of consumption is the percentage change in welfare derived from a percentage change in consumption or income.

Figure 5. Mean Panelist Ratings of the Importance of Categories of Uncertainties Associated with Benefit and Cost Estimates from Integrated Assessment Models

The Majority of Panelists Said That the United States Should Begin to Control Emissions Soon, Regardless of International Participation

The majority of panelists said that if the United States acts unilaterally to establish a price on greenhouse gas emissions, they would not change their conclusions that their recommended actions were economically justified. While some panelists said that the United States should proceed cautiously if it acts unilaterally, 16 of 18 panelists agreed that the United States should establish a price on greenhouse gas emissions as soon as possible, regardless of the extent to which other countries adopt similar policies. Nonetheless, the majority of the panelists said that it was important for the United States to participate in international negotiations to facilitate climate agreements or to enhance the credibility or influence of the United States. In addition, some panelists noted the importance of participation by other countries, recommending that the United States act conservatively in the absence of action by other high-emitting nations. For example, one panelist said that the U.S. government should also engage in international negotiations to ensure other nations make a similar commitment. In emphasizing the importance of global participation, the panelist said that without it, a U.S. emissions reduction program would be undercut.

When asked whether U.S. action to establish a price on greenhouse gas emissions in the absence of action by other high-emitting nations (e.g., India, China, Brazil) would have a negative or positive effect on the ability of U.S. companies to compete with similar companies in other countries, 10 out of 18 panelists said the effect would be negative, 7 said it would be neither positive or negative, and 1 did not know or was not sure.

Moreover, several panelists said that energy-intensive industries such as chemicals and metals, as well as the coal sector, would experience the most negative effects if the United States established a price on greenhouse gases in the absence of such actions in other high-emitting nations.

On the other hand, one panelist remarked that the major impacts would not necessarily come from international trade, adding that the main effect of U.S. action would be on electricity production, which is not generally traded internationally. Another panelist said that competitiveness is only one factor if the United States takes action to mitigate greenhouse gas emissions and the other key countries do not. The most important factor, this panelist noted, is that in the absence of action by other high-emitters a large investment in abatement of emissions by the United States will be wasted because it will not achieve a corresponding improvement in the environment.

The Panelists' Views on the Strengths and Limitations of Policy Options Focused Primarily on the Environmental Certainty of a Cap-and-Trade System versus the Efficiency of a Tax on Emissions

The panel provided their opinions on the strengths and limitations associated with various policy options to address climate change, and focused on the key trade-offs between a tax on emissions or a cap-andtrade system. Panelists first identified criteria for evaluating policy options and then discussed the strengths and limitations of different policy options within this context. The most important trade-offs identified by panelists focused on the environmental effectiveness of a cap-and-trade system versus the economic efficiency of a tax on emissions. While panelists viewed other policy options less favorably, they cited their potential as a complement to a market-based mechanism. These expert opinions should be of assistance to the Congress in weighing the potential benefits and costs of different policies for addressing climate change.

Panelists Rated the Importance of Criteria for Evaluating Policy Options

In our first questionnaire, panelists identified criteria that they believe the Congress should consider in evaluating the various actions and policy options for addressing climate change. In the second questionnaire, panelists rated the importance of those criteria, such as economic efficiency and environmental effectiveness. Economic efficiency is used to assess whether a policy alternative would maximize net benefits (that is, where marginal benefits equal marginal costs) to society, and environmental effectiveness means that the policy implemented has the desired environmental result. The criteria were divided into two categories: efficiency-related criteria, such as economic efficiency, environmental effectiveness, cost effectiveness, flexibility and adaptability of policies; and equity-related criteria, such as political feasibility, impact on international negotiations, and the distribution of costs and benefits among generations and countries. On average, the panelists rated cost effectiveness and political feasibility as the most important criteria (see Figure 6 and app. III for more detail).

The Most Important Tradeoffs are between the Relative Effectiveness of Cap-and-Trade Systems and the Relative Efficiency of Taxes

When asked to identify their preferred policy options, 8 of the 18 panelists said that they prefer a cap and trade with a safety valve as a way to combine some elements of both a cap-and-trade system and a tax. Some panelists noted that a traditional cap-and-trade system would provide more environmental certainty; that is, it would more likely achieve specific emissions reductions because the policy caps the total amount of emissions at a specific level. On the other hand, several panelists said a tax on emissions would be more economically efficient, citing, for example, its ability to provide certainty about marginal abatement costs associated with controlling greenhouse gases and avoiding price volatility that could occur in a permit market. In addition, some noted that a cap-and-trade program can be more administratively burdensome than a tax. Nonetheless, some of the panelists that preferred a tax said that a cap-andtrade program, especially if it included cost minimizing components, would be an acceptable second option to address climate change.

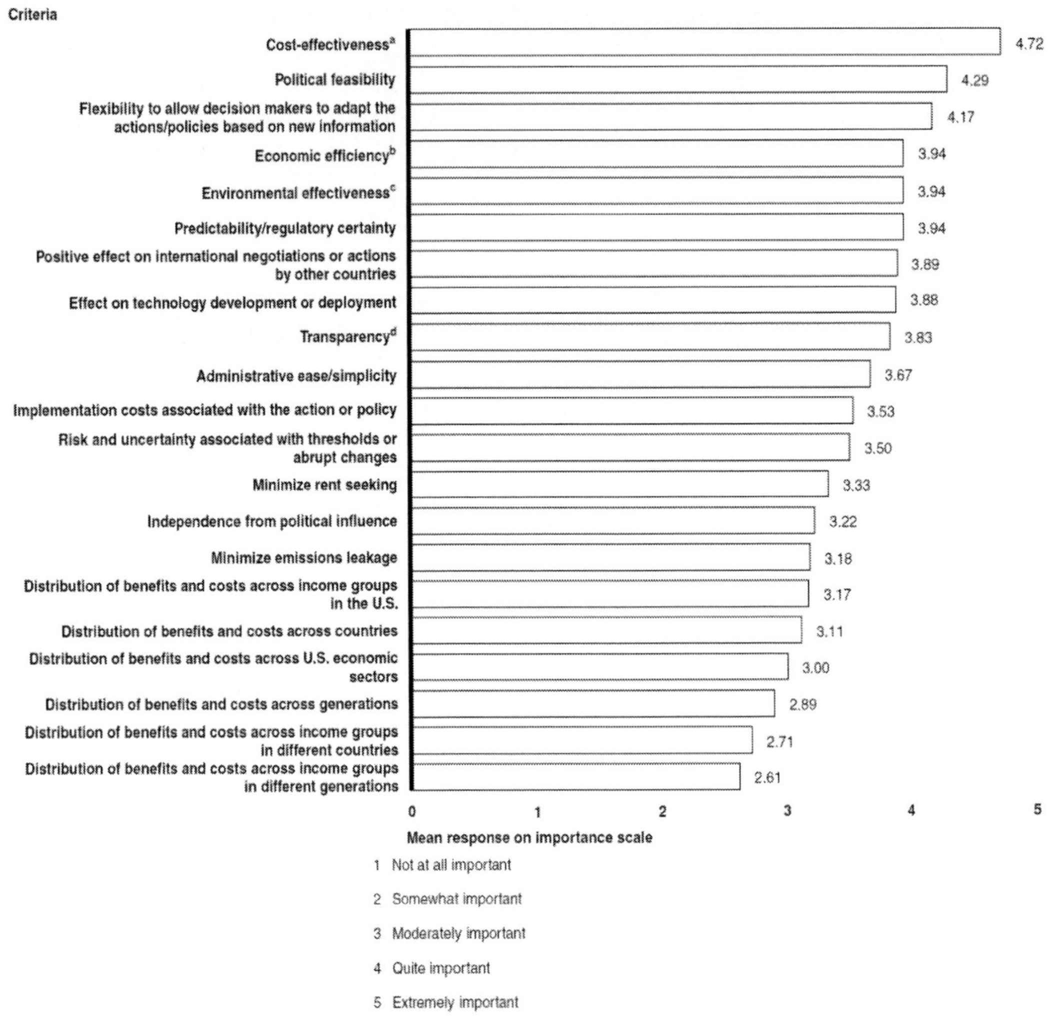

Source: GAO analysis.

Note: Total responses for each question may range from 15 to 18. See appendix III for additional details.

[a] Cost-effectiveness is the extent to which a goal is achieved in a least cost manner.

[b] Economic efficiency occurs when, from a societal perspective, production is at an efficient level (e.g. net benefits are maximized).

[c] Environmental effectiveness is the extent to which policy achieves the environmental target.

[d] Transparency is the extent to which policy's requirements and costs are visible to all parties.

Figure 6. Mean Panelist Ratings of the Importance of Criteria in Evaluating Policies for Addressing Climate Change

The majority of panelists supporting a safety valve cited the potential for limiting volatility as an important rationale for incorporating a safety valve into a cap-and-trade program. In addition, some panelists cited uncertainty regarding the marginal costs and marginal benefits of complying with a cap-and-trade program, and some cited the flexibility that it would provide to decision makers to respond to new information as rationales for having a safety valve. However, 3 panelists expressed concern regarding the compatibility of a safety valve with an international greenhouse gas trading system. For example, one panelist stated that under any realistic safety valve system, the permits should be good only in the country of issue. Otherwise, countries with higher market emissions prices could buy permits in the United States, undermining their respective emissions targets. Another panelist added that if the United States commits to specific emissions targets as part of an international agreement, a safety valve may affect the United States' ability to meet those commitments. If the safety valve is triggered, for example, emitters will be allowed to purchase additional permits to emit above the targeted cap for emissions, which may keep the United States from meeting agreed-to goals. Potential solutions offered by panelists included abolishing the safety valve after the emissions market has stabilized and then integrating into an international system, and incorporating a quantitatively limited safety valve where a limited number of permits were sold once the safety valve was reached.

A majority of the panelists said that the government should auction off rather than give away at least a portion of the permits under a cap-andtrade system. The means by which the government distributes allowances can have important economic implications. For example, the existing cap-and-trade programs for sulfur dioxide (a pollutant that causes acid rain) and for carbon dioxide in the European Union give away most of the allowances to regulated entities with a limited number of allowances reserved for auction. The majority of panelists believed that the government should use either a combination of auctioning and free allocations or auctioning to distribute the allowances. In addition, we asked the panelists to rate the importance of various actions to distribute the revenue generated from auctioning allowances that could offset adverse effects on consumers or particular sectors of the economy. On average, the panelists rated using the revenues to reduce the tax burden for low-income individuals as the most important way to distribute the revenue (see Figure 7 and app. III for more detail). In addition, on average, panelists rated using revenues to support research and development in various areas, such as zero- or low-carbon technologies and carbon capture and storage as at least moderately important.

The Panelists Viewed Other Policy Options Less Favorably but Cited Their Potential as a Complement to a Market-Based Mechanism

In the second questionnaire, we asked panelists to discuss the strengths and weaknesses of other domestic policy options that would serve to complement a tax or cap-and-trade system, but some viewed them less favorably in the absence of a market-based mechanism. In weighing the strengths and weaknesses of policy options, we asked panelists to discuss research and development of technologies, adaptation to climate change, revising efficiency standards, and reforming subsidies. For example, several panelists noted that mechanisms such as vehicle fuel efficiency standards would be unnecessary if the Congress enacted a mitigation policy to place a price on carbon. Two panelists noted that setting efficiency standards for vehicles was unlikely to yield cost-effective reductions in emissions because greater fuel efficiency lowers the cost of driving, potentially leading individuals to drive more. Table 3 reflects examples cited by the panelists of strengths and limitations of specific policy options.

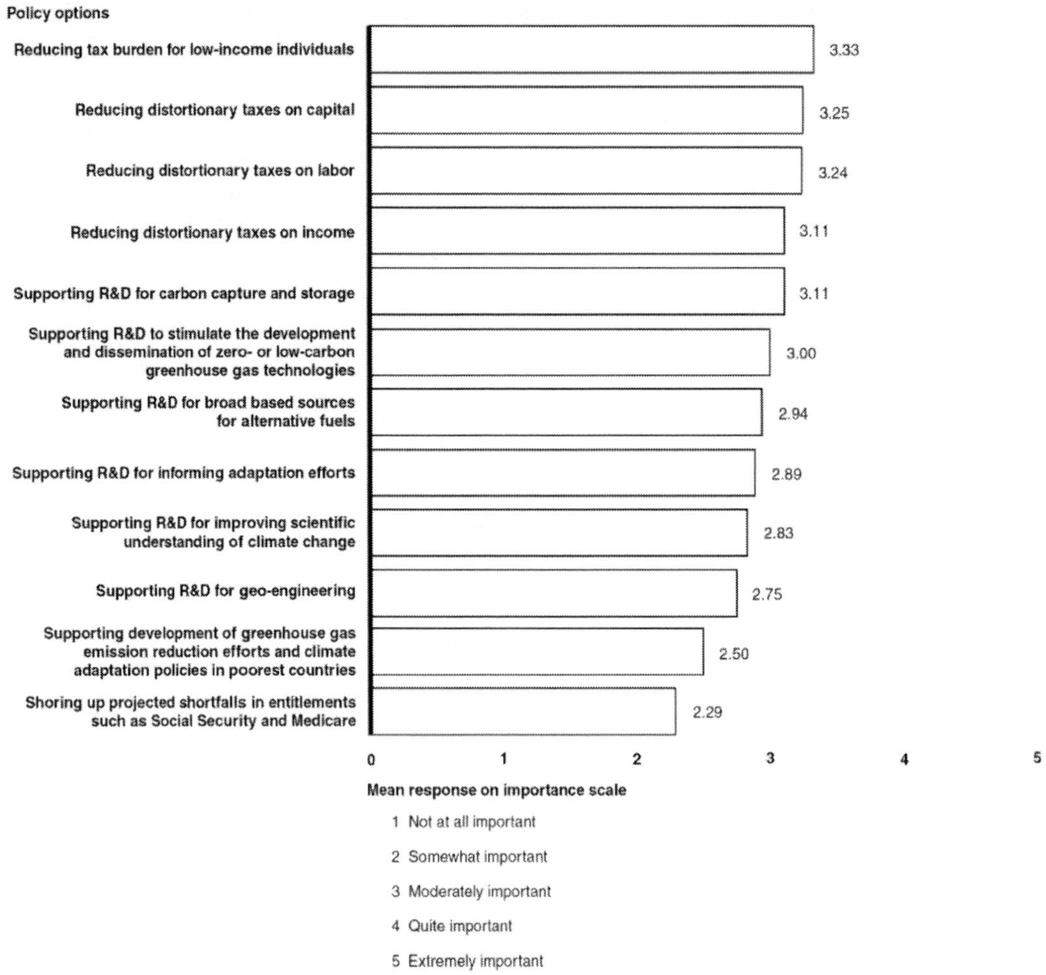

Source: GAO analysis.

Note: Total responses for each question may range from 15 to 18. See appendix III for additional details.

Figure 7. Mean Panelist Ratings of the Importance of Ways That Revenue Generated from a Market-Based Mechanism Could Be Distributed

The panelists cited the strengths and weaknesses of public investments in the development and deployment of technologies that could reduce emissions. Some suggested such efforts could complement a market-based mitigation policy by, for example, lowering the costs of controlling emissions and enhancing the likelihood of achieving long-term emissions targets, while others noted that public investment may be unnecessary with a mitigation policy in place. In addition, some panelists said that while they supported research and development investment, it was not enough on its own without additional policies to address climate change. Further, several panelists also expressed concern about the government picking what it perceives as "winning" technologies rather than funding a broader research and development program.

Table 3. Key Strengths and Limitations of Other Policy Options Identified by Panelists

Policy option	Examples of strengths	Examples of limitations
R&D in low- or zero-carbon technologies	• May help lower long-term costs of mitigation • Promotes the development of new technologies and encourages innovation • Facilitates international technology transfer • Government-funded R&D can be disseminated quickly • Privately funded R&D may be slow to develop	• Insufficient as a policy on its own • Unnecessary in the presence of a robust mitigation policy • Uncertainty regarding feasibility, costs, and timing of deployment • Danger of funding less useful technologies due to politics or uncertainty • Could crowd out other sources of R&D funding
Adaptation	• Essential to help federal, state, and local governments plan and prepare for unavoidable consequences • Clear role for the federal government in some sectors, such as water resources and land management	• Insufficient policy on its own • Difficult to define • Private sector has incentives to adapt without federal intervention • Federal government has little role
		• Will yield only moderate payoffs in terms of reducing future costs • Removes incentive to reduce emissions or develop better energy technologies
Reforming subsidies for fuels or energy-efficient technologies	• Necessary to make current policies more efficient • Facilitates the development of better fuels and energy-efficient technologies • May result in better pricing of resources	• Insufficient policy on its own • Unnecessary in the presence of a robust mitigation policy • Inferior to mitigation for addressing climate change • Potential for rent seeking[a] • Economically inefficient • Politically difficult
Develop or revise efficiency standards	• Could reduce the overall costs of mitigation • Important in the presence of real market failures, such as buildings and consumer appliances • Politically feasible	• Insufficient policy on its own • Unnecessary in the presence of a robust mitigation policy • Economically inefficient • Unequal burden of costs • Limited potential benefits
R&D on the basic science of climate change	• Informs adaptation needs • Guides stringency of climate policy goals • Reduces uncertainty regarding future impacts, costs, and benefits of climate change • Enhances monitoring	• Insufficient policy on its own • Unnecessary in the presence of a robust mitigation policy • Sufficient knowledge on the subject • Adequately funded at current levels • Could postpone action on mitigation • Difficult to determine the proper amount of R&D needed

Source: GAO analysis of panelists' responses

[a] Rent seeking occurs when a party (e.g., individual or organization) seeks an economic gain (e.g., subsidy) from the government.

A majority of panelists said that funding efforts to adapt to the impacts of climate change were at least moderately important. Some panelists identified actions the federal government could take to prepare for climate change impacts such as reforming insurance subsidy programs in areas vulnerable to natural disasters like hurricanes or flooding and creating capacity within federal agencies to protect against and react to climate change impacts. When discussing the strengths and weaknesses of actions to adapt to climate change, several panelists noted that implementing adaptation policies may help reduce vulnerabilities when faced with inevitable climate change or in the event of possible catastrophic climate change. One panelist referred to adaptation as a risk management strategy, or form of insurance, against uncertain future events resulting from climate change. Some panelists noted that a clear federal role exists for certain sectors, such as water resource management and property rights protection, which could require additional resources for infrastructure development, research, and adapting practices to use alternative methods for distributing water and managing federal lands. Another panelist said that implementing adaptation policies could also help ensure greater international and economic equity, since some areas likely to be affected by climate change are underdeveloped, economically disadvantaged, or vulnerable in some way to impacts. However, some panelists stated that adaptation was difficult to define, making the federal role unclear, and others said that incentives for adaptation may already exist, limiting the need for a federally directed adaptation policy.

Panelists identified the strengths and limitations of other policy options to address climate change, such as subsidy reform for alternative fuels or energy-efficient technologies and research and development into the basic science of climate change. For example, in rating the importance of additional actions for the Congress to consider, 12 panelists rated reforming subsidies for alternative fuels or energy-efficient technologies as at least moderately important, and several panelists noted that subsidies for alternative fuels or energy-efficient technologies are economically inefficient or insufficient on their own without an appropriate mitigation policy. Regarding research and development on the basic science of climate change, several panelists said that it could help inform adaptation efforts and reduce uncertainty of costs or benefits by improving understanding of the potential dangers from climate change.[30] However, some panelists noted that additional funding of research and development on the basic science of climate change is not warranted, citing reasons such as a belief that it is adequately funded at current levels, it could be used as an excuse to delay taking action to mitigate climate change, and that the priority should instead be placed on adaptation efforts and monitoring.

As agreed with your offices, unless you publicly announce the contents of this report earlier, we plan no further distribution until 30 days from the report date. At that time, we will send copies to House Committees on Energy and Commerce; Natural Resources; Science and Technology; Transportation and Infrastructure; and the Select Committee on Energy Independence and Global Warming. We will also provide the report to the Senate Committees on Environment and Public Works; Energy and Natural Resources; Commerce, Science and Transportation; and the Joint Economic Committee. Copies will be made available to others on request, and the report will be available at no charge on the GAO Web site at http://www.gao.gov.

If you or your staff have any questions regarding this report, please contact me at (202) 512-3841 orstephensonj@gao.gov. Contact points for our Offices of Congressional Relations and Public Affairs may be found on the last page of this report. Key contributors are listed in appendix VI.

John B. Stephenson
Director, Natural Resources and Environment

APPENDIX I. SCOPE AND METHODOLOGY

To address the first and second objectives, we (1) reviewed relevant climate change academic literature and documents developed by federal agencies, (2) met with agency officials from the Environmental Protection Agency (EPA), the Department of Energy (DOE) including the Energy Information Administration (EIA), the Department of Commerce including the National Oceanic and Atmospheric Administration (NOAA), the United States Department of Agriculture (USDA), the National Aeronautic and Space Administration (NASA), and the Council of Economic Advisors (CEA), and (3) obtained expert opinion on the economic effects of actions to address climate change and the strengths and weaknesses of those actions using a virtual panel on the Internet.

To structure and gather expert opinions from the panel, we employed a modified version of the Delphi method. The method is based on a structured process for collecting and distilling knowledge from a group of experts by means of a series of questionnaires.[31] Used to support informed decision making, the Delphi method was first developed at the RAND Corporation in the 1950s. One of the strengths of this approach is its flexibility, and while first used in a live group discussion format, the method is easily modified for various settings. The modified process we employed utilized two Web-based questionnaires, and incorporated an iterative and controlled feedback process to gather the experts' opinions. Specifically, experts' responses to the first questionnaire were used to create the questions for the second, allowing the experts to consider the opinions and issues raised by other panelists when responding to the second round of questions. Also, by using a Web-based process, we were able to overcome some of the potential biases associated with group discussions. These biasing effects include the potential dominance of individuals and group pressure for conformity. Moreover, by creating a virtual panel, we were able to include more experts than would have been possible with a live panel. While the method has these strengths, there are some potential limitations. For example, there is considerable reliance on the active participation of the panelists, which can vary widely, and some panelists may not complete the entire study. In addition, the results of the iterative process are limited to the issues, topics, and responses generated by those participating; thus some topics or viewpoints may not be considered in the process. To mitigate the latter limitation, we added critical questions to the second questionnaire that were not discussed in the first round based on our review of the literature.

We contracted with the National Academy of Sciences (NAS) to select and recruit a panel of experts with a range of in-depth experience assessing the economic impacts of climate change policy. Participants were to (1) have expertise in modeling and analyzing benefits, costs, and uncertainties using integrated assessment methods; (2) have knowledge of policies for mitigating climate change; (3) have knowledge of the economic trade-offs associated with different policies for mitigating climate change; and (4) be affiliated with U.S.-based institutions including academia, the federal government, and other research-oriented entities. To select the experts, we provided NAS with a preliminary list of potential panelists that we identified in our review of the literature. Taking these names into account, NAS developed a list of 37 who met our criteria. We reviewed and agreed with the list of names, and NAS sent 25 individuals an electronic letter via e-mail inviting them to participate in the study along with a description of the project. Invitees were given the option of discussing the project further with a project representative before deciding whether or not to participate, which some chose to do. Of the 25 panelists NAS recruited to participate, 21 agreed and were sent the first questionnaire. Nineteen responded to the first questionnaire, and 18 responded to the second.[32] All of the experts who participated completed a form stating that they had no conflicts of interest that would compromise their ability to participate in the panel.

Prior to the posting of the questionnaires, we conducted a series of pretests with internal and external experts, including two panel participants. The goals of the pretests were to check that (1) the questions were clear and unambiguous and (2) terminology was used correctly. We made changes to the content and format of both questionnaires as necessary during the pretesting processes. We also conducted usability tests of both questionnaires for the Internet to ensure operability. For each phase of the Delphi, we posted a questionnaire on the Internet. Panel members were notified of the availability of the questionnaire with an e-mail message. The e-mail message contained a unique user name and password that allowed each respondent to log on and fill out a questionnaire but did not allow respondents access to the questionnaires of others.

In the first phase, we asked panelists to provide responses to three open- ended questions on the economics of actions to address climate change developed from an extensive literature review. We asked the panelists to (1) identify what actions, if any, the Congress should consider to address climate change; (2) provide estimates of the benefits and costs of their recommended actions; and (3) identify criteria that are appropriate for evaluating the potential actions. In addition, we asked the panelists to provide citations to support their responses.

After the first questionnaire was completed, we performed a content analysis of the open-ended responses. Using the themes and topics that were discussed, as well as the varying opinions of the panel, the second questionnaire was constructed. For example, panelists were asked to rate the importance of the various assumptions and uncertainties associated with integrated assessment models that the experts collectively identified in the first questionnaire. Using this approach, panelists could provide their opinions on the topics that others had raised, and areas of agreement and disagreement could be identified. In addition, this approach allowed for the panelists to reevaluate their original responses in light of the responses of the whole group. While these first round responses were the primary source for developing the second questionnaire, literature cited in support of their responses, as well as information gathered from GAO's literature review were also incorporated as necessary. The second phase questionnaire included mostly closed-ended questions, with a limited number of open-ended questions, and encapsulated panelists' views on preferred policy options, potential benefits and costs, key uncertainties, and the strengths and weaknesses of different policy options.

Panel members had approximately 4 weeks between July and August of 2007 to complete their questionnaires in the first phase of the panel, approximately 4 weeks in October and November of 2007 to complete their questionnaires in the second phase of the panel, and approximately 4 weeks to provide comment (at their discretion) on a summary of their second round responses in January of 2008. Selected questions and aggregated responses from the second phase are presented in appendix III. In addition, we asked several follow-up questions requesting that panelists clarify their responses or elaborate on critical policy issues. While we display only the quantitative, closed-ended responses, we also relied on the responses to the qualitative, open-ended questions to inform our findings in this report. GAO provided a summary of the findings of this report and briefed representatives from the CEA, the Council on Environmental Quality (CEQ), EPA, and DOE on the results of the panel prior to issuing this report. The views expressed by the panel members do not necessarily represent the views of GAO.

APPENDIX II. SELECTED CHARACTERISTICS OF PANELISTS' PREFERRED POLICY OPTIONS FOR ADDRESSING CLIMATE CHANGE

Listed below are selected panelist policy recommendations for implementing a market-based mechanism to place a price on greenhouse gas emissions.

Table 4. Selected Characteristics of Panelists' Preferred Policy Options for Addressing Climate Change

Panelist	Preferred market-based policy	Greenhouse gases covered	Date of implementation	Scope	Point of regulation[a]	Initial price range [b]	Safety valve pricec	Complementary policy actions
Panelist 1	Cap and trade	All gases	2010-2015	Economywide	Combination	> 14	N/A	(1) R&D on low carbon energy technologies; (2) energy efficiency and renewable portfolio standards
Panelist 2	Cap and trade	All gases	2010-2015	Economywide	Upstream	0.3-3	N/A	Not specified
Panelist 3	Cap and trade	All gases	2010-2015	Economywide	Combination	21-30	N/A	(1) R&D subsidies; (2) adaptation
Panelist 4	Cap and trade w/ safety valve	Carbon dioxide	2010-2015	Economywide	Upstream	21-30	Don't know	Not specified
Panelist 5	Cap and trade w/ safety valve	All gases	Before 2010	Economywide	Upstream	0.3-3	0.55	(1) International negotiation and assistance
Panelist 6	Cap and trade w/ safety valve	All gases	2010-2015	Economywide	Downstream	3-5	3.81	(1) Do away with import tariffs and agricultural subsidies; (2) tax breaks for private sector research and development; (3) remove perverse subsidies; (4) declassify energy saving and lightweight technologies for export to China and India
Panelist 7	Cap and trade w/ safety valve	Carbon dioxide	Before 2010	Economywide	Combination	3-5	5	(1) Funding R&D; (2) funding adaptation; (3) reform policies that discourage energy efficiency

Table 4. (Continued)

Panelist	Preferred market-based policy	Greenhouse gases covered	Date of implementation	Scope	Point of regulationa	Initial price range b	Safety valve pricec	Complementary policy actions
Panelist 8	Cap and trade w/ safety valve	All gases	2010-2015	Economywide	Upstream	11-20	3 to 4 times the initial price of emissions	(1) Negotiate international agree-ments with major greenhouse gas emitters; (2) invest in pilot demon-stration projects for carbon capture and storage and renewable fuels.
Panelist 9	Cap and trade w/ safety valve	All gases	2010-2015	Economywide	Upstream	11-20	12 to 20	(1) Support for technology research, development, and deployment; (2) support for developing countries; (3) international engagement to leverage our action for action by other countries.
Panelist 10	Cap and trade w/ safety valve	All gases	Before 2010	Economywide including land use	Combination	1-10	2.70	(1) Limit loopholes that could arise in policies that focus on specific sectors or groups of economic actors.
Panelist 11	Cap and trade w/ safety valve	All gases	Before 2010	Economywide	Downstream	11-20	25	(1) Research and development; (2) technology transfer (both empha-sizing adaptation and emissions reduction strategies).
Panelist 12	Tax	Carbon dioxide	Before 2010	Economywide	Upstream	11-20	N/A	(1) Support carbon capture and storage; (2) sponsor research and development in alternative energy sources; (3) sponsor adaptation at home and abroad.
Panelist 13	Tax	All gases	Before 2010	Economywide	Downstream	0.3-3	N/A	Not specified
Panelist 14	Tax	All gases	Before 2010	Economywide	Combination	6-8	N/A	(1) Basic research in the underlying science for low-carbon energy systems.

Table 4. (Continued)

Panelist	Preferred market-based policy	Greenhouse gases covered	Date of implementation	Scope	Point of regulation[a]	Initial price range[b]	Safety valve price[c]	Complementary policy actions
Panelist 15	Tax	Carbon dioxide first, then expand to all	2010-2015	Economywide	Upstream	11-20	N/A	(1) Adaptation; (2) R&D (including geo-engineering)
Panelist 16	Tax	All gases	2010-2015	Economywide incorporating carbon sinks from land use	Mostly upstream, midstream for utilities and carbon capture and storage	11-20	N/A	(1) Independently directed effort to support R&D; (2) international assistance and negotiations to facilitate action by other nations
Panelist 17	Tax	No answer	Before 2010	Economywide	Downstream	41-50	N/A	Not specified
Panelist 18	Tax	All gases	Before 2010	Economywide	Upstream	N/A[d]	N/A	Not specified

Source: GAO.

Note: N/A stands for not applicable.

[a] Point of regulation in the economy. Combination means the policy would have both upstream and downstream provisions.

[b] Initial price range per metric ton of carbon dioxide equivalent in 2007 dollars.

[c] Safety valve price per metric ton of carbon dioxide equivalent in 2007 dollars.

[d] One panelist preferred a policy that set a price for each individual greenhouse gas as opposed to one price for all emissions.

APPENDIX III: SELECTED QUESTIONS AND EXPERT RESPONSES

Listed below are questions and summary panelist responses supporting figures 2 through 7 of this report. Panelists were asked to rate the importance of various items related to policy approaches and rationales for addressing climate change (Mean responses were calculated by assigning the following values to the rated level of importance: 1 = not at all, 2 = somewhat, 3 = moderately, 4 = quite, and 5 = extremely).

Question 1: In addition to establishing a price on anthropogenic sources of emissions using either a tax or cap and trade system (or other regulatory approach), panelists identified the following list of actions that Congress might also consider as part of a broader portfolio of actions. How important is it that Congress consider each action to address climate change?

	Panelist responses							
	Rated level of importance							
Additional actions to address climate change	Not at all	Somewhat	Moderately	Quite	Extremely	Don't know/ no response	Number of responses	Mean level of importance
Participating in international negotiations to facilitate participation in climate agreements or to enhance U.S. credibility/ influence	0	1	2	2	12	1	17	4.47
Funding R&D and/or deployment of low- or zero-carbon technologies	1	5	2	5	5	0	18	3.44
Creating an independent oversight board to establish, monitor, and revise the mechanism for controlling greenhouse gases (tax or cap and trade)	3	3	1	4	5	2	16	3.31
Providing international assistance to developing countries	1	5	5	3	4	0	18	3.22
Reforming subsidies for alternative fuels or energy-efficient technologies	1	4	5	5	2	1	17	3.18
Funding efforts to adapt to impacts of climate change by reforming agricultural subsidy programs, land use practices, flood control, etc.	1	5	4	6	2	0	18	3.17
Developing or revising technology efficiency standards (for fuels or energy use)	6	1	3	6	1	1	17	2.71
Funding private sector R&D to achieve cost-effective emissions reductions in developing countries	4	7	3	3	1	0	18	2.44
Funding R&D on domestic and international public health impacts that could arise as a result of climate change	2	12	1	2	1	0	18	2.33
Levying a graduated tax on cars and light trucks (including sport utility vehicles), based on fuel economy	7	3	5	2	1	0	18	2.28

Source: GAO.

Question 2: In the first round of questions, panelists identified criteria that the Congress should consider in evaluating the various actions and policy options for addressing climate change. The list below summarizes these criteria. How important would you say that the following criteria are in evaluating the various policies for addressing climate change?

Panelist responses								
	Rated level of importance							
Criteria	Not at all	Somewhat	Moderately	Quite	Extremely	Don't know / no response	Number of responses	Mean level of imortance
Cost-effectiveness	0	0	0	5	13	0	18	4.72
Political feasibility	0	1	1	7	8	1	17	4.29
Flexibility to allow decision makers to adapt the actions/policies based on new information	0	0	4	7	7	0	18	4.17
Economic efficiency	0	1	5	6	6	0	18	3.94
Environmental effectiveness	0	2	3	7	6	0	18	3.94
Predictability/regulatory certainty information	0	2	4	5	7	0	18	3.94
Positive effect on international negotia-tions or actions by other countries	0	3	1	9	5	0	18	3.89
Effect on technology development or deployment	0	1	4	8	4	1	17	3.88
Transparency	0	2	3	9	4	0	18	3.83
Administrative ease/simplicity	0	2	4	10	2	0	18	3.67
Implementation costs associated with the action or policy	0	3	4	8	2	1	17	3.53
Risk and uncertainty associated with thresholds or abrupt changes	1	3	3	8	3	0	18	3.50
Minimize rent seeking	2	2	6	4	4	0	18	3.33
Independence from political influence	1	4	6	4	3	0	18	3.22
Minimize emissions leakage	1	3	6	6	1	1	17	3.18
Distribution of benefits and costs across income groups in the United States	0	4	8	5	1	0	18	3.17
Distribution of benefits and costs across countries	0	3	11	3	1	0	18	3.11
Distribution of benefits and costs across U.S. economic sectors	0	7	5	5	1	0	18	3.00
Distribution of benefits and costs across generations	1	7	5	3	2	0	18	2.89
Distribution of benefits and costs across income groups in different countries	0	9	4	4	0	1	17	2.71
Distribution of benefits and costs across income groups in different generations	1	10	2	5	0	0	18	2.61

Source: GAO.

Question 3: Panelists identified several actions that Congress should consider in deciding how to distribute the revenue that would be generated from a federal program to establish a price on greenhouse gases (from an imposition of a tax or from an auction of permits). How important is it for Congress to distribute the generated revenue to each of these categories?

	Panelist responses							
	Rated level of importance							
Uses for revenue distribution	**Not at all**	**Somewhat**	**Moderately**	**Quite**	**Extremely**	**Don't know /no rsponse**	**Number of responses**	**Mean level of importance**
Reducing tax burden for low-income individuals	1	4	5	4	4	0	18	3.33
Reducing distortionary taxes on capital	1	1	8	5	1	2	16	3.25
Reducing distortionary taxes on labor	0	3	9	3	2	1	17	3.24
Reducing distortionary taxes on income	0	4	9	4	1	0	18	3.11
Supporting R&D for carbon capture and storage	2	4	6	2	4	0	18	3.11
Supporting R&D to stimulate the development and dissemination of zero- or low-carbon/greenhouse gas technologies	2	4	5	6	1	0	18	3.00
Supporting R&D for broad-based sources for alternative fuels	3	3	7	2	3	0	18	2.94
Supporting R&D for informing adaptation efforts	3	5	4	3	3	0	18	2.89
Supporting R&D for improving scientific understanding of climate change	2	6	5	3	2	0	18	2.83
Supporting R&D for geo engineering	3	4	5	2	2	2	16	2.75
Supporting development of green house gas emission reduction efforts and climate adaptation policies in poorest countries	2	9	4	2	1	0	18	2.50
Shoring up projected shortfalls in entitlements such as Social Security and Medicare	6	5	2	3	1	1	17	2.29

Source: GAO.

Question 4: Many panelists said that it is difficult to estimate the potential benefits associated with emissions mitigation partly because of uncertainty about the impact of climate change and the actions that could be used to address it. In some cases, the panelists identified general categories of benefits that could accrue from various policy options. These categories are below. How important are the following categories of potential benefits as a rationale for addressing climate change?

Panelist responses								
	Rated level of importance							
Categories of benefits of actions to address climate change	**Not at all**	**Somewhat**	**Moderately**	**Quite**	**Extremely**	**Don't know / no response**	**Number of responses**	**Mean level of importance**
Avoided climate change damages (including flooding, extreme weather events, crop damage, impacts on sensitive ecosystems, public health, and species loss)	0	0	1	11	5	1	17	4.24
Reducing risk or the likelihood of extreme events associated with climate change	1	0	2	8	6	1	17	4.06
Establishing a price signal to influence market or individual behavior	3	1	0	2	10	2	16	3.94
Facilitation of international cooperation on climate change	3	1	0	5	7	2	16	3.75
Development of a regulatory system for greenhouse gas emissions	3	2	2	4	5	2	16	3.38
Inducing investment in developing countries	4	3	4	1	3	3	15	2.73
Development, deployment, and, more generally, creating incentives for businesses and consumers to invest in low- or zero-carbon technologies	5	3	1	6	1	2	16	2.69
Reducing risk of international conflict over natural resources and territory	4	2	7	2	1	2	16	2.63

Source: GAO.

Question 5: Panelists identified several key assumptions that they made in estimating the benefits and costs using integrated assessment models. How important are the following assumptions in terms of affecting the estimated benefits and costs in integrated assessment models?

Panelist responses								
	Rated level of importance							
Key assumptions	Not at all	Somewhat	Moderately	Quite	Extremely	Don't know / no response	Number of responses	Mean level of importance
Real discount rate for assessing benefits and costs of addressing climate change	0	1	2	3	11	1	17	4.41
Inclusion of all economic sectors	0	1	1	9	6	1	17	4.18
Baseline economic growth	0	0	2	11	3	2	16	4.06
Degree of international participation	0	2	1	8	6	1	17	4.06
Climate sensitivity	0	0	3	10	4	1	17	4.06
Baseline energy use	0	0	3	11	3	1	17	4.00
Rate of technological change	0	1	3	9	3	2	16	3.88
Likelihood of abrupt change or extreme events	0	2	4	7	4	1	17	3.76
Energy price projections	0	2	5	8	2	1	17	3.59
Price elasticities	0	0	7	9	0	2	16	3.56
Coverage across greenhouse gases	0	2	7	5	3	1	17	3.53
Baseline population growth	0	1	10	5	1	1	17	3.35
Baseline adaptation	1	4	4	6	2	1	17	3.24
Efficient use of revenue by reducing distortionary taxes	0	5	7	4	1	1	17	3.06
Ancillary public health benefits	1	3	12	1	0	1	17	2.76

Source: GAO.

Question 6: Numerous respondents identified the importance of considering risk and uncertainty in identifying policy options for the Congress to consider. Below is a list of the key uncertainties that panelists identified. In general, how important are each of these uncertainties in terms of affecting the estimated benefits and costs in integrated assessment models?

Panelist responses								
	Rated level of importance							
Key uncertainties	Not at all	Somewhat	Moderatel y	Quite	Extremely	Don't know/ no response	Number of responses	Mean level of importance
Thresholds and abrupt changes in climate system	0	1	4	5	8	0	18	4.11
Science of climate change	0	0	5	8	5	0	18	4.00
Economic effect of actions to address climate change	0	2	4	8	4	0	18	3.78
Discounting (in general)	1	4	2	5	6	0	18	3.61
How people will adapt to climate change	0	2	6	8	2	0	18	3.56
Pure rate of time preference	1	4	3	5	5	0	18	3.50
Elasticity of the marginal utility of consumption	2	4	4	6	1	1	17	3.00

Source: GAO.

APPENDIX IV. PANEL OF EXPERTS

Joseph Aldy, Resources for the Future
James Edmonds, Pacific Northwest National Laboratory
Richard Howarth, Dartmouth College
Bruce McCarl, Texas A&M University
Robert Mendelsohn, Yale University
William Nordhaus, Yale University
Sergey Paltsev, Massachusetts Institute of Technology
William Pizer, Resources for the Future
David Popp, Syracuse University
John Reilly, Massachusetts Institute of Technology
Roger Sedjo, Resources for the Future
Kathleen Segerson, University of Connecticut
Brent Sohngen, Ohio State University
Robert Stavins, Harvard University
Richard Tol, Economic and Social Research Institute
Martin Weitzman, Harvard University
Peter Wilcoxen, Syracuse University
Gary Yohe, Wesleyan University

APPENDIX V. BIBLIOGRAPHY OF SELECTED LITERATURE REVIEWED BY GAO

Bullets identify citations provided by panelists in their responses.

Aldy, Joseph E; Barrett Scott, and Stavins Robert N. "Thirteen Plus One: A Comparison of Global Climate Policy Architectures." *Climate Policy*, vol. 3 (2003): 373-397.

- Ahmed; Rasha and Kathleen Segerson. *Emissions Control and the Regulation of Product Markets: The Case of Automobiles*. Working Paper, Department of Economics, University of Connecticut: July 2007.
- Antle, JM; Capalbo, SM; Paustian, K and Ali, MK. "Estimating the Economic Potential for Agricultural Soil Carbon Sequestration in the Central United States Using an Aggregate Econometric-Process Simulation Model," *Climatic Change*, vol. 80 (2007): 145-171.Ashenfelter, Orley, and Karl Storchmann. Using a Hedonic Model of Solar Radiation to Assess the Economic Effect of Climate Change: The Case of Mosel Valley Vineyards. Working Paper 12380, National Bureau of Economic Research. Cambridge: July 2006.

Azar, Christian, and Lindgren, Kristian "Catastrophic Events and Stochastic Cost-Benefit Analysis of Climate Change: Editorial." *Climatic Change*, vol. 56, (2003): 245-255.

- Brown, PG. "Toward an Economics of Stewardship: The Case of Climate." *Ecological Economics*, Vol. 26 (1998): 11-21.

- Burtraw et al. "Ancillary Benefits of Reduced Air Pollution in the United States from Moderate Greenhouse Gas Mitigation Policies in the Electricity Sector." *Journal of Environmental Economics and Management*, 45 (2003): 650-73.
- United States Climate Change Science Program. *Scenarios of Greenhouse Gas Emissions and Atmospheric Concentrations*. 2007.
- Cline, William R. *The Economics of Global Warming*. Washington, D.C.: Institute for International Economics, 1992.

Congressional Budget Office. *Evaluating the Role of Prices and R&D in Reducing Carbon Dioxide Emissions*. Pub. 2731. Washington, D.C.: September 2006.

Congressional Budget Office. *Trade-offs in Allocating Allowances for Carbon Dioxide Emissions*. Washington, D.C.: April 25, 2007.

Congressional Budget Office. *Uncertainty in Analyzing Climate Change: Policy Implications*. Washington, D.C.: January 2005.

Congressional Research Service. *Climate Change: Design Approaches for a Greenhouse Gas Reduction Program*. RL33799. Washington, D.C.: Updated January 16, 2007.

Congressional Research Service. *U.S. Global Climate Change Policy: Evolving Views on Cost, Competitiveness, and Comprehensiveness*. RL30024. Washington, D.C.: Updated January 29, 2007.

DeCanio; Stephen J; Howarth, Richard B; Sanstad, Alan H; Schneider, Stephen H and Thompson, Starley L. *New Directions in the Economics and Integrated Assessment of Global Climate Change*. Arlington, Va.: Pew Center on Global Climate Change: October 2000.

Edmonds, Jae; Clarke, John; Dooley, James; Kim, Son H and Smith, Steven J "Stabilization of Carbon Dioxide in a B2 world: Insights on the Roles of Carbon Capture and Disposal, Hydrogen, and Transportation Technologies." *Energy Economics*, vol. 26, (July 2004): 517-537.

Edmonds, Jae; Wise, Marshall and Barns, David W "Carbon Coalitions: The Cost and Effectiveness of Energy Agreements to Alter Trajectories of Atmospheric Carbon Dioxide Emissions." *Energy Policy*, vol. 23, no. 4/5 (1995): 309-335.

Energy Information Administration. *Energy Market and Economic Impacts of a Proposal to Reduce Greenhouse Gas Intensity with a Cap and Trade System*. SR/OIAF/2007-01. Washington, D.C.: January, 2007.

Environmental Protection Agency. Tools of the Trade: A Guide to Designing and Operating a Cap and Trade Program for Pollution Control. EPA430-B-03-002. Washington, D.C.: June, 2003.

Fankhauser, Samuel. *Valuing Climate Change: The Economics of the Greenhouse*. London, UK: Earthscan Publications Limited, 1995.

Fankhauser, Samuel and Tol, Richard SJ. "On Climate Change and Economic Growth." *Resource and Energy Economics*, vol. 27, (2005): 1-17.

- Fischer, Carolyn and Morgenstern, Richard D. "Carbon Abatement Costs: Why the Wide Range of Estimates?" *The Energy Journal*, vol. 27, no. 2 (2006):73-86.
- Fischer, Carolyn and Newell, Richard "Environmental and Technology Policies for Climate Mitigation." *Resources for the Future Discussion Paper*, RFF DP 04-05, February 2007.

Gerlagh, Reyer and Zwaan, Bob van der "Options and Instruments for a Deep Cut in Carbon Dioxide Emissions: Carbon Dioxide Capture or Renewables, Taxes or Subsidies?" *The Energy Journal*, vol. 27, no. 3 (2006).

Gurgel, Angelo C; Paltsev, Sergey; Reilly, John M and Metcalf, Gilbert E. *U.S. Greenhouse Gas Cap-and Trade Proposals: Application of a Forward-Looking Computable General Equilibrium Model.* Report No. 150. Cambridge, Mass.: MIT Joint Program on the Science and Policy of Global Change, June 2007.

Heal, Geoffrey and Kristrom, Bengt "Uncertainty and Climate Change." *Environmental and Resource Economics*, vol. 22 (2002): 3-39.

Helm, Dieter. "The Assessment: Climate Change Policy." *Oxford Review of Economic Policy*, vol. 19, no. 3 (November 2003):349-361.

Hope, Chris. "The Marginal Impact of CO2 from PAGE2002: An Integrated Assessment Model Incorporating the IPCC's Five Reasons for Concern." *The Integrated Assessment Journal*, vol. 6, iss. 1 (2006): 19-56.

- Howarth, RB. "Against High Discount Rates." In *Perspectives on Climate Change Science, Economics, Politics, Ethics*, edited by W. Sinnott- Armstrong and R. B. Howarth. Amsterdam: Elsevier, 2005.

Intergovernmental Panel on Climate Change. *Climate Change 1995: Economic and Social Dimensions of Climate Change*. Contribution of Working Group III to the Second Assessment Report of the Intergovernmental Panel on Climate Change. Cambridge, UK: Cambridge University Press, 1995.

- Intergovernmental Panel on Climate Change. "Summary for Policymakers," in *Climate Change 2001: The Scientific Basis*. Cambridge, UK: Cambridge University Press, 2001.
- Intergovernmental Panel on Climate Change. "Summary for Policymakers," in *Climate Change 2001: Impacts, Adaptation, and Vulnerability*. Cambridge, UK: Cambridge University Press, 2001.
- Intergovernmental Panel on Climate Change. "Summary for Policymakers," in *Climate Change 2007: The Physical Science Basis*. Cambridge, UK: Cambridge University Press, 2007.

Intergovernmental Panel on Climate Change. *Climate change 2007: Climate Change Impacts, Adaptation and Vulnerability*. Cambridge, UK: Cambridge University Press, 2007.

- Intergovernmental Panel on Climate Change. *Climate change 2007: Mitigation of Climate Change*. Cambridge, UK: Cambridge University Press, 2007.
- Interlaboratory Working Group on Energy-Efficient and Clean Energy Technologies. *Scenarios for a Clean Energy Future*. Oak Ridge, Tenn., and Berkeley, Calif.: Oak Ridge National Laboratory and Lawrence Berkeley National Laboratory, 2000.
- Jaffe, Adam B; Newell, Richard G and Stavins, Robert N. "Technology Policy for Energy and the Environment." In *Innovation Policy and the Economy*, vol 4, edited by Adam B. Jaffe, Josh Lerner, and Scott Stern. MIT Press: Cambridge, Mass., pp. 35-68, 2004.
- Jones, R and Yohe, GW. (2007), "Applying Risk-Analytic Techniques to the Integrated Assessment of Climate Policy Benefits," Produced for the Global Forum

on Sustainable Development on the Economic Benefits of Climate Change Policies, July 2006, Paris, France.

Jorgenson, Dale W; Goettle, Richard J; Hurd, Brian H and Smith, Joel B. *U.S. Market Consequences of Global Climate Change*. Arlington, Va.: Pew Center on Global Climate Change, 2004.

- Kammen, Daniel M and Nemet, Gregory F. "Reversing the Incredible Shrinking Energy R&D Budget." *Issues in Science and Technology*, 22 (2005): 84-88.

Kavuncu, Y. Okan and Knabb, Shawn D. "Stabilizing Greenhouse Gas Emissions: Assessing the Intergenerational Costs and Benefits of the Kyoto Protocol." *Energy Economics*, vol. 27 (2005):369-386.

Keller; Klaus; Bolker, Benjamin M and Bradford, David F. "Uncertain Climate Thresholds and Optimal Economic Growth." *Journal of Environmental Economics and Management*, vol. 48 (2004): 723-74 1.

Keller, Klaus; Hall, Matt; Kim, Seung-Rae; Bradford, David F and Oppenheimer, M. "Avoiding Dangerous Anthropogenic Interference with the Climate System." *Climatic Change*, vol. 73 (2005): 227-238.

Kelly, David L and Kolstad, Charles D. "Integrated Assessment Models for Climate Change Control." In *International Yearbook of Environmental and Resource Economics 1 999/2000: A Survey of Current Issues*, edited by Henk Folmer and Tom Tietenberg, pages 171-197. Cheltenham, UK: Edward Elgar, 1999.

- Knetsch, JL. "Gains, Losses, and the US-EPA Economic Analyses Guidelines: A Hazardous Product?" *Environmental and Resource Economics* 32 (2005): 91-1 12.

Kolstad, Charles and Toman, Michael *The Economics of Climate Policy*. Discussion Paper 00-40REV. Washington, D.C.: Resources for the Future, June 2001.

Lange, Andreas; Vogt, Carsten and Ziegler, Andreas. "On the Importance of Equity in International Climate Policy: An Empirical Analysis." *Energy Economics*, vol. 29 (2007): 545-562.

Lewandrowski, Jan; Peters, Mark; Jones, Carol; House, Robert; Sperow, Mark; Eve, Marlen and Paustian, Keith. *Economics of Sequestering Carbon in the U.S. Agricultural Sector*. Technical Bulletin TB 1909. Washington, D.C.: U.S. Department of Agriculture Economic Research Service, March, 2004.

Lind, Robert C. "Intergenerational Equity, Discounting, and the Role of Cost-Benefit Analysis in Evaluating Global Climate Policy." *Energy Policy*, vol. 23, no. 4/5 (1995): 379-389.

Maddison, David. "A Cost-Benefit Analysis of Slowing Climate Change." *Energy Policy*, vol. 23, no. 4/5 (1995): 337-346.

Manne, Alan; Mendelsohn, Robert and Richels, Richard "MERGE : A Model for Evaluating Regional and Global Effects of GHG Reduction Policies." *Energy Policy*, vol. 23, (1995): 17-34.

Manne, Alan and Richels, Richard "The Impact of Learning-by-Doing on the Timing and Costs of CO2 Abatement." *Energy Economics*, vol. 26, issue 4 (2004): 603-619.

- Manne, Alan and Richels, Richard. "The Kyoto Protocol: A Cost-Effective Strategy for Meeting Environmental Objectives?" *The Energy Journal* 20 (1999): 1-23.
- Mastrandrea, MD and Schneider, SH. "Probabilistic Integrated Assessment of 'Dangerous' Climate Change." *Science* 304 (2004): 571-575

- McCarl, BA and Sands, RD. "Competitiveness of Terrestrial Greenhouse Gas Offsets: Are They a Bridge to the Future?" *Climatic Change* 80 (2007): 109-126.

McFarland, JR; Reilly, JM and Herzog, HJ. "Representing Energy Technologies in Top-down Economic Models Using Bottom-up Information." *Energy Economics*, vol. 26 (2004): 685-707.

- McKibbin, Warwick J and Wilcoxen, Peter J. *Climate Change Policy after Kyoto: Blueprint for a Realistic Approach.* Washington, D.C.: Brookings Institution Press, 2002.
- McKibbin, Warwick J and Wilcoxen, Peter J. "A Credible Foundation for Long Term International Cooperation on Climate Change." In *Architectures for Agreement: Addressing Global Climate Change in the Post-Kyoto World*, edited by Joseph Aldy and Robert Stavins. Cambridge University Press, 2007.
- McKibbin, Warwick J and Wilcoxen, Peter J. "The Role of Economics in Climate Change Policy." *Journal of Economic Perspectives*, vol. 16, no. 2 (2002): 107-129.

Mendelsohn, Robert. "Efficient Adaptation to Climate Change." *Climatic Change*, vol. 45 (2000): 583-600.

Mendelsohn, Robert. "The Role of Markets and Governments in Helping Society Adapt to a Changing Climate." *Climatic Change*, vol. 78 (2006): 203-2 15.

Montgomery, David W and Smith, Anne E. *Price, Quantity, and Technology Strategies for Climate Change Policy.* Washington, D.C.: CRA International, October 11, 2005.

Morgan, Granger M and Dowlatabadi, Hadi "Learning from Integrated Assessment of Climate Change." *Climatic Change*, vol. 34 (1996): 337-368.

Neumann, James; Yohe, Gary; Nicholls, Robert and Manion, Michelle *Sea- Level Rise and Global Climate Change: A Review of Impacts to U.S. Coasts*. Arlington, Va.: Pew Center on Global Climate Change, February, 2000.

Newell, Richard G; Jaffe, Adam B and Stavins, Robert N. "The Effects of Economic and Policy Incentives on Carbon Mitigation Technologies." *Energy Economics*, vol. 28 (2006): 563-578.

Newell, Richard and Pizer, William "Discounting the Distant Future: How Much Do Uncertain Rates Increase Valuations?" Discussion paper 00-45. Washington, D.C.: Resources for the Future, 2000.

- Newell, R; Pizer, W and Zhang, J. "Managing Permit Markets to Stabilize Prices." *Environmental and Resource Economics* 31 (2005): 133-157.
- Nordhaus, William D. "The Challenge of Global Warming: Economic Models and Environmental Policy." New Haven, Conn.: Yale University, July 24, 2007.
- Nordhaus, William D. "Critical Assumptions in the Stern Review on Climate Change." *Science*, vol. 317 (July 13, 2007): 201-202.
- Nordhaus, William D. "Economic Analyses of Kyoto Protocol: Is There Life After Kyoto?" Conference paper presented at Global Warming: Looking Beyond Kyoto, Yale University, New Haven, Conn., October 2 1-22, 2005.
- Nordhaus, William D. "Expert Opinion on Climatic Change." *American Scientist*, vol. 82. (1994): 45-51.

- Nordhaus, William D. "Geography and Macroeconomics: New Data and New Findings." *Proceedings of the National Academy of Sciences*, 103 (2006): 3510-3517.

Nordhaus, William D. Managing the Global Commons: The Economics of Climate Change. Cambridge: The MIT Press, 1994.

Nordhaus, William D. "An Optimal Transition Path for Controlling Greenhouse Gases." Science, vol. 258 (1992): 1315-1319.

Nordhaus, William D. *The "Stern Review" on the Economics of Climate Change*. Working Paper 12741. Cambridge: National Bureau of Economic Research, December 2006.

Nordhaus, William D. "To Slow or Not to Slow: The Economics of the Greenhouse Effect." *The Economic Journal*, vol. 101 (1991): 920-937.

- Nordhaus William D. "To Tax or Not to Tax? Alternative Approaches to Slowing Global Warming." *Review of Environmental Economics and Policy* 1(1) (Winter 2007): 26-44.
- Nordhaus, William D and Boyer, Joseph *Warming the World: Economic Models of Global Warming*. Cambridge: the MIT Press, 2000.

Organisation for Economic Co-operation and Development. *The Benefits of Climate Change Policies*. Paris, France: 2004.

- Pacala, S and Socolow, R. "Stabilization Wedges: Solving the Climate Problem for the Next 50 Years with Current Technologies." *Science* 305 (2004): 968-972.
- Paltsev, S; Reilly, J; Jacoby, H; Gurgel, A; Metcalf, G; Sokolov, A and Holak, J. "Assessment of US Cap-and-Trade Proposals." MIT Joint Program on the Science and Policy of Global Change, Report 146 (2007).

UK Parliament House of Lords Select Committee on Economic Affairs. *The Economics of Climate Change*. Volume I: Report. HL-12-I. London, UK: 2005.

UK Parliament House of Lords Select Committee on Economic Affairs. *The Economics of Climate Change*. Vol. 2: Evidence. HL-12-II. London, UK: 2005.

UK Parliament House of Lords Select Committee on Economic Affairs. *Government Response to the Economics of Climate Change*. HL-71. London, UK: 2005.

Pearce, David. "The Political Economy of an Energy Tax: The United Kingdom's Climate Change Levy." *Energy Economics*, vol. 28 (2006): 149- 158.

Pearce, David. "The Role of Carbon Taxes in Adjusting to Global Warming." *The Economic Journal*, vol. 101, no. 407 (1991): 938-948.

Pearce, David. "The Social Cost of Carbon and its Policy Implications." *Oxford Review of Economic Policy*, vol. 19, no. 3 (2003): 362-384.

Peck, Stephen C and Teisberg, Thomas J. "Optimal Carbon Emissions Trajectories when Damages Depend on the Rate or Level of Global Warming." *Climatic Change*, vol. 28 (1994): 289-314.

Pindyck, Robert S. *Uncertainty in Environmental Economics*. Washington, D.C.: AEI-Brookings Joint Center for Regulatory Studies, December 2006.

- Pizer, William A. "Combining Price and Quantity Controls to Mitigate Global Climate Change." *Journal of Public Economics* 85 (2002): 409-434.
- Pizer, William A. "The Evolution of a Global Climate Change Agreement." *American Economic Review* 96(2) (2006): 26-30.

Pizer, William A. "The Optimal Choice of Climate Change Policy in the Presence of Uncertainty." *Resource and Energy Economics*, vol. 21 (1999): 255-287.

Pizer, William A; Burtraw, Dallas; Harrington, Winston; Newell, Richard and Sanchirico, James *Modeling Economywide versus Sectoral Climate Policies Using Combined Aggregate-Sectoral Models.* Discussion Paper 05-08. Washington, D.C.: Resources for the Future, April, 2003.

- Popp, David. "R&D Subsidies and Climate Policy: Is There a 'Free Lunch'?" *Climatic Change* 77(3-4) (2006): 311-341.

Portney, Paul R and Weyant, John P. Editors. *Discounting and Intergenerational Equity*. Washington D.C.: Resources for the Future, 1999.

Reilly, John M. Editor. Agriculture: The Potential Consequences of Climate Variability and Change for the United States. Cambridge, UK: Cambridge University Press, 2002.

- Reilly, J and Asadoorian, M. "Mitigation of Greenhouse Gas Emissions from Land Use: Creating Incentives Within Greenhouse Gas Emissions Trading Systems." *Climatic Change* 80 (2007): 173-197.
- Reilly J; Paltsev, S; Felzer, B; Wang, X; Kicklighter, D; Melillo, J; Prinn, R; Sarofim, M; Sokolov, A and Wang, C. "Global Economic Effects of Changes in Crops, Pasture, and Forests Due to Changing Climate, Carbon Dioxide, and Ozone." MIT Joint Program on the Science and Policy of Global Change, Report No. 149, May 2007.
- Reilly, J; Sarofim, M; Paltsev, S and Prinn, R. "The Role of Non-CO2 Greenhouse Gases in Climate Policy: Analysis Using the MIT IGSM." *Energy Journal*, Special Issue on Multigas Mitigation and Climate Policy (2006): 503-520.

Riahi, Keywan; Rubin, Edward S; Taylor, Margaret R; Schrattenholzer, Leo and Hounshell, David "Technological Learning for Carbon Capture and Sequestration Technologies." *Energy Economics*, vol. 26, (2004): 539-564.

Richels, Richard and Edmonds, Jae "The Economics of Stabilizing Atmospheric CO2 Concentrations." *Energy Policy*, vol. 23, no. 4/5 (1995): 373-378.

Richels, Richard; Rutherford, Thomas; Blanford, Geoffrey and Clarke, Leon *Managing the Transition to Climate Stabilization.* Working Paper 07-01. Washington, D.C.: AEI-Brookings Joint Center for Regulatory Affairs, January 2007.

Schelling, Thomas C. "Intergenerational Discounting." *Energy Policy*, vol. 23, no. 4/5 (1995): 395-401.

Schneider, Stephen and Lane, Janica "Integrated Assessment Modeling of Global Climate Change: Much Has Been Learned—Still a Long and Bumpy Road Ahead." *The Integrated Assessment Journal*, vol. 5, iss. 1 (2005): 41- 75.

Smith, Joel B. *A Synthesis of Potential Climate Change Impacts* on the U.S. Arlington, Va.: Pew Center on Global Climate Change, April, 2004.

Sohngen, Brent; Sedjo, Roger; Mendelsohn, Robert and Lyon, Ken. *Analyzing the Economic Impact of Climate Change on Global Timber Markets*. Discussion Paper 96-08. Washington, D.C.: Resources for the Future, 1996.

- Stavins, Robert N. "Proposal for a U.S. Cap-and-Trade System to Address Global Climate Change: A Sensible and Practical Approach to Reduce Greenhouse Gas

Emissions." The Hamilton Project, the Brookings Institution, Washington, D.C.: September 12, 2007.

Stavins, Robert N; Jaffe, Judson and Schatzki, Todd *Too Good to be True? An Examination of Three Economic Assessments of California Climate Change Policy*. Related Publication 07-01. Washington, D.C.: AEI- Brookings Joint Center for Regulatory Affairs, January, 2007.

Stavins, Robert N and Richards, Kenneth R. *The Cost of U.S. Forest- Based Carbon Sequestration*. Arlington, Va.: Pew Center on Global Climate Change, 2005.

Stern, Nicholas. *The Economics of Climate Change*: The Stern Review. Cambridge, UK: Cambridge Press, 2007.

Sutherland, Ronald J. "'No Cost' Efforts to Reduce Carbon Emissions in the U.S.: An Economic Perspective." *Energy Journal*, vol. 21, no. 3 (2000): 89-1 12.

- Tavoni, M; Sohngen, Brent and Bosetti, Valentina "Forestry and Carbon Market Response to Stabilize Climate." *Energy Policy*, 35 (2007), 5346– 5353.
- Thomas, CD; et al. "Extinction Risk from Climate Change." *Nature* 247 (2004): 145-148.

Tol, Richard SJ. "The Damage Costs of Climate Change Toward More Comprehensive Calculations." *Environmental and Resource Economics*, vol. 5 (1995): 353-374.

Tol, Richard SJ. "Europe's Long-Term Climate Target: A Critical Evaluation." *Energy Policy*, vol. 35, no. 1 (2007): 424-432.

Tol, Richard SJ. "Is the Uncertainty About Climate Change Too Large for Expected Cost-Benefit Analysis?" *Climatic Change*, vol. 56 (2003): 265- 289.

Tol, Richard SJ. "The Marginal Costs of Greenhouse Gas Emissions." *The Energy Journal*, vol. 20, no. 1 (1999): 61-81.

- Tol, Richard SJ. "The Marginal Damage Costs of Carbon Dioxide Emissions: An Assessment of the Uncertainties." *Energy Policy*, 33 (2005): 2064-2074.

Tol, Richard SJ. "Welfare Specifications and Optimal Control of Climate Change: An Application of FUND." *Energy Economics*, vol. 24 (2002): 367- 376.

Tol, Richard SJ and Yohe, Gary "A Review of the Stern Review." *World Economics*, vol. 7, no. 4 (2006): 233-250.

Van Vuuren, Detlef; Weyant, John and Chesnaye, Francisco de la. "Multi- Gas Scenarios to Stabilize Radiative Forcing." *Energy Economics*, vol. 28 (2006): 102-120.

Webster, Mort D; Babiker, Mustafa H; Mayer, Monika; Reilly, John M; Harnisch, Jochen; Hyman, Robert; Sarofim, Marcus C and Wang, Chien *Uncertainty in Emissions Projections for Climate Models*. Report No. 79. Cambridge: MIT Joint Program on the Science and Policy of Global Change, August, 2001.

- Webster, Mort; Forest, C; Reilly, J; Babiker, M; Kicklighter, D; Mayer, M; Prinn, R; Sarofim, M; Sokolov, A; Stone, P and Wang, C. "Uncertainty Analysis of Climate Change and Policy Response." *Climatic Change*, 61 (2003): 295-320.

Weitzman, Martin L. "Prices versus Quantities." *The Review of Economic Studies*, vol. 41, no. 4 (1974): 477-491.

- Weitzman, Martin L. "The Role of Uncertainty in the Economics of Catastrophic Climate Change." AEI-Brookings Joint Center for Regulatory Studies, Working Paper 07-11, May 2007.

Weitzman, Martin L. *Structural Uncertainty and the Value of Statistical Life in the Economics of Catastrophic Climate Change*. Working Paper 07-11. Washington, D.C.: AEI-Brookings Joint Center for Regulatory Studies, Revised October 2007.

Weyant, John P. "Introduction and Overview." *Energy Economics*, vol. 26 (2004): 501-515.

Wigley, TML. "The Kyoto Protocol: CO2, CH4 and Climate Implications." *Geophysical Research Letters* 25 (1998): 2285–2288.

Wigley, TML; Richels, R and Edmonds, JA. "Economic and Environmental Choices in the Stabilization of Atmospheric CO2 Concentrations." *Nature*, vol. 379 (1996): 240-243.

Wing, Ian Sue. "Representing Induced Technological Change in Models for Climate Policy Analysis." *Energy Economics*, vol. 28 (2006): 539-562.

- Yohe, G. "Carbon Emissions Taxes: Their Comparative Advantage under Uncertainty." *Annual Review of Energy* 17 (1992): 301-326.
- Yohe, G. "More on the Properties of a Tax Cum Subsidy Pollution Control Strategy." *Economic Letters* 31 (1989): 193-198.

Yohe, Gary. "Some Thoughts on the Damage Estimates Presented in the Stern Review—An Editorial." *The Integrated Assessment Journal,* vol. 6, iss. 3 (2006): 65-72.

- Yohe, Gary; Andronova, Natasha and Schlesinger, Michael "To Hedge or Not Against an Uncertain Climate Future?" *Science*, vol. 306 (2004): 416- 417.

Yohe, Gary and Schlesinger, Michael E. "Sea-level change: The Expected Economic Cost of Protection or Abandonment in the United States." *Climatic Change*, vol. 38 (1998): 447-472.

Yohe, G; Schlesinger, ME and Andronova, NG. "Reducing the Risk of a Collapse of the Atlantic Thermohaline Circulation." *The Integrated Assessment Journal*, vol. 6 (2006): 57-73.

- Yohe, GW; Tol, RSJ and Murphy, D. "On Setting Near-term Climate Policy while the Dust Begins to Settle: The Legacy of the Stern Review." *Energy and Environment* 18 (2007): 621-633.

END NOTES

[1] Major greenhouse gases include carbon dioxide (CO_2), methane (CH_4), nitrous oxide (N_2O), and synthetic gases (hydrofluorocarbons {HFCs}, perfluorocarbons {PFCs}, and sulfur hexafluoride {SF_6}).

[2] See GAO,*Climate Change: Financial Risks to Federal and Private Insurers in Coming Decades Are Potentially Significant,* GAO-07-285, Mar. 16, 2007; and *Climate Change: Agencies Should Develop Guidance for Addressing the Effects on Federal Land and Water Resources*, GAO-07-863, (Washington, D.C.: Aug. 7, 2007).

[3] Atmospheric concentrations of carbon dioxide increased from 280 parts per million to 379 parts per million between pre-industrial times and 2005. (See Intergovernmental Panel on Climate Change (IPCC), 2007: The Physical Science Basis. Contribution of Working Group I to the Fourth Assessment)

[4] Environmental Protection Agency (EPA), 2008, *Inventory of U.S. Greenhouse Gas Emissions and Sinks, 1990-2006,* Public Review Draft (Washington, D.C.: Apr. 18, 2008).

[5] Energy Information Administration (EIA), *International Energy Annual 2005.*

[6] In the context of the United Nation's Framework Convention on Climate Change, mitigation is a human intervention to reduce the sources or enhance the sinks of greenhouse gases. Examples include using fossil fuels more efficiently for industrial processes or electricity generation, switching to solar energy or wind power, improving the insulation of buildings, and expanding forests and other sinks to remove greater amounts of carbon dioxide from the atmosphere. Forests and other vegetation are considered sinks because they remove carbon dioxide through photosynthesis.

[7] In general, auctioning the allowances would enable the government to decide how to use the revenue, and allocating the allowances for free would represent a transfer of wealth from the government to the entities receiving the allowances.

[8] In general, a firm will purchase permits when the permit price is lower than the cost to abate emissions.

[9] Establishing a price on emissions creates a price signal to fossil fuel users to cut back on consumption.

[10] National Oceanic and Atmospheric Administration (NOAA), "Coastal Area and Marine Resources: The Potential Consequences of Climate Variability and Change," December 2001.

[11] The appropriate price, often called the social cost of carbon, reflects the present value of economic damages caused by an additional quantity of greenhouse gas emissions. Under an economically optimal control policy, the price would be set at the point where the incremental or marginal damages from global warming equal the marginal costs of controlling emissions.

[12] International Panel on Climate Change, Working Group II (2007). A metric ton is equivalent to 1,000 kilograms, or approximately 2,204 pounds.

[13] One panelist preferred a policy that set a price for each individual greenhouse gas as opposed to one price for all emissions.

[14] Because greenhouse gases differ in their potential to contribute to global warming, each gas is assigned a unique weight, called a global warming potential, based on its heat- absorbing ability relative to carbon dioxide over a fixed period. This provides a way to convert emissions of various greenhouse gases into a common measure, called carbon dioxide equivalent.

[15] IPCC, 2007: Introduction. *In: Climate Change 2007: Mitigation, Contribution of Working Group III to the Fourth Assessment Report of the Intergovernmental Panel on Climate Change (B.* Metz, O.R. Davidson, P.R. Bosch, R. Dave, L.A. Meyer [eds.]), Cambridge University Press, Cambridge, United Kingdom.

[16] IPCC, 2007: Introduction. In: *Climate Change 2007: Mitigation, Contribution of Working Group III to the Fourth Assessment Report of the Intergovernmental Panel on Climate Change* (B. Metz, O.R. Davidson, P.R. Bosch, R. Dave, L.A. Meyer [eds.]), Cambridge University Press, Cambridge, United Kingdom.

[17] The IPCC defines mitigation as technological change and substitution that reduce resource inputs, such as energy use, and emissions per unit of output. Although several social, economic, and technological policies would produce an emissions reduction, with respect to climate change, mitigation means implementing policies to reduce greenhouse gas emissions and enhance greenhouse gas sinks.

[18] UNFCCC, Bali Action Plan, December 2007.

[19] National Research Council: *Abrupt Climate Change: Inevitable Surprises* (Washington, D.C.: 2002).

[20] In general, benefits and costs are measured in terms of a common metric: dollars. Benefits are monetized by estimating the amount that individuals would be willing to pay for the benefit (or the amount they would be willing to accept to forgo the benefit) based on market transactions. In cases where the benefit is not traded in a market (e.g., ecosystem services), economists use methods such as contingent valuation surveys to elicit the amount that individuals would be willing to pay for the benefit.

[21] A recent report sponsored by the British government (i.e., the Stern Review) concluded that relatively aggressive emissions reductions by the global community were economically justified. Several economists criticized its discounting approach. See appendix V for literature citations on this topic.

[22] IPCC, 2007: Summary for Policymakers. In *Climate Change 2007: Impacts, Adaptation and Vulnerability. Contribution of Working Group II to the Fourth Assessment Report of the Intergovernmental Panel on Climate Change*, M. L. Parry, O. F. Conziani, J. P. Palutikof, P. J. van der Linden, and C. E. Hanson, Eds., Cambridge University Press, Cambridge, United Kingdom, 7-22.

[23] S. 280 (Lieberman), S. 309 (Sanders), S. 317 (Feinstein), S. 485 (Kerry), S. 1766 (Bingaman), S. 2191 (Lieberman/Warner), H.R. 620 (Olver), and H.R. 1590 (Waxman).

[24] The remaining 4 panelists did not provide a response.

[25] In specifying an initial price for a cap-and-trade or hybrid program, the emissions cap could be set at a level that would be consistent with the initial price (i.e., the targeted permit price).

[26] One panelist preferred a policy that set a price for each individual greenhouse gas as opposed to one price for all emissions.

[27] Several noted that some categories of benefits, such as establishing a price signal, are not benefits per se because they are not the result of reducing emissions. Nonetheless, because these categories were identified by other panelists as benefits of actions to address climate change, we asked all the panelists to rate their importance.

[28] CCSP cited (1) the amount that emissions must be reduced to achieve an emissions path to stabilization, and (2) the technologies that are available to facilitate changes in the economy as reasons for the difference in stabilization costs between models.

[29] The science of climate change refers to knowledge about the effect on the climate of factors such as greenhouse gas concentrations, clouds, and the carbon cycle.

[30] The science of climate change refers to knowledge about the effect on the climate of factors such as greenhouse gas concentrations, clouds, and the carbon cycle.

[31] Adler, Michael, and Eric Ziglio, eds. *Gazing into the Oracle: The Delphi Method and Its Application to Social Policy and Public Health* (Bristol, Pennsylvania: 1996).

[32] The responses presented in this report are from the second questionnaire, supplemented with examples and anecdotes from the open-ended responses in the first round, and represent the views of the 18 panelists who participated in both phases of the panel.

In: Climate Change Measures and Trade Considerations ISBN: 978-1-60876-756-4
Editor: George R. Fried

Chapter 2

CLIMATE CHANGE LEGISLATION: INTERNATIONAL TRADE CONSIDERATIONS

Eileen Claussen

Mr. Chairman, Mr. Grassley, members of the Committee, thank you for the opportunity to testify on the international trade considerations of climate change legislation. My name is Eileen Claussen, and I am the President of the Pew Center on Global Climate Change.

The Pew Center on Global Climate Change[1] is an independent non-profit, non-partisan organization dedicated to advancing practical and effective solutions and policies to address global climate change. Our work is informed by our Business Environmental Leadership Council (BELC), a group of 44 major companies, most in the Fortune 500, that work with the Center to educate opinion leaders on climate change risks, challenges, and solutions. The Pew Center is also a founding member of the U. S. Climate Action Partnership, a coalition of 25 leading businesses and five environmental organizations that have come together to call on the federal government to quickly enact strong national legislation to require significant reductions of greenhouse gas emissions.

Addressing global climate change presents policy challenges at both the domestic and the international levels, and the issue of competitiveness underscores the very close nexus between the two. The immediate task before this Committee, and before the Senate, is developing and enacting a comprehensive domestic program to limit and reduce U.S. greenhouse gas (GHG) emissions. Moving forward with a mandatory program to reduce U.S. emissions in advance of a comprehensive international agreement presents both risks and opportunities. On the one hand, domestic GHG limits may lead to a shift of some energy-intensive production to countries without climate constraints, resulting in "emissions leakage" and posing competitiveness concerns for some domestic industries. On the other hand, a mandatory domestic program in the United States is an essential step towards the development of an effective global climate agreement.

In the long term, a strong multilateral framework ensuring that all major economies contribute their fair share to the global climate effort is, I believe, the most effective means of addressing competitiveness concerns. Achieving such an agreement must be a fundamental objective of U.S. climate policy. In designing a domestic climate program, the question before Congress is what to do about the potential for leakage in the interim – until an effective global agreement is in place. In considering this question, it is important to distinguish two distinct but closely related policy challenges: (1) how best to encourage strong climate action by other countries, and in particular, by the major emerging economies; and (2) how best to minimize potential competitiveness impacts on U.S. industry. I believe that each of these two objectives is most effectively addressed through a different set of policy responses, and it is important to ensure that our efforts to address one do not undermine the other.

I will focus today primarily on the second of these challenges: designing transitional policies to minimize potential competitiveness impacts on U.S. industry.[2] Our analysis of the underlying issues leads us to conclude that the potential competitiveness impacts of domestic climate policy are modest and are manageable.

In my testimony, I will:

1. present our analysis of the nature and potential magnitude of the competitiveness challenge;
2. discuss a range of options for addressing competitiveness concerns; and
3. outline what we believe would be the most effective approach. This approach would employ output-based emission allocations to vulnerable industries, phased out over time, and other transition assistance to affected workers and communities.

Understanding Competitiveness Concerns

A first step in considering options to address competitiveness is assessing the potential scope and magnitude of potential competitiveness impacts. It is important to note that it is not the competitiveness of the U.S. economy as a whole that is at issue. (According to the Environmental Protection Agency's (EPA) analysis of the American Clean Energy and Security (ACES) Act of 2009 passed last month by the House, the cost of meeting the bill's emission reduction targets in 2030 would be a 0.37 percent loss in GDP.[3] Put another way, GDP would reach $22.6 trillion, nearly 60 percent higher than today, approximately two months later than without the bill.) Rather, the concern centers on a relatively narrow segment of the U.S. economy: energy-intensive industries whose goods are traded globally, such as steel, aluminum, cement, paper, glass, and chemicals. As heavy users of energy, these industries will face higher costs as a result of domestic GHG constraints; however, as the prices of their goods are set globally, their ability to pass along these price increases is limited.

Competitiveness impacts can be experienced as a loss in market share to foreign producers, a shift in new investment, or, in extreme cases, the relocation of manufacturing facilities overseas. In assessing the economic consequences of past environmental regulation in the United States, most analyses find little evidence of significant competitive harm to U.S. firms. Many studies conclude that other factors—such as labor costs, the availability of capital, and proximity to raw materials and markets—weigh far more heavily in firms' location decisions. One comprehensive review—synthesizing dozens of studies of the impact of U.S. environmental regulation on a range of sectors—concluded that while new environmental rules imposed significant costs on regulated industries, they did not appreciably affect patterns of trade.[4]

In the case of GHG regulation, the additional cost to firms could include the compliance cost of purchasing allowances to cover direct emissions; indirect compliance costs embedded in higher fuel or electricity prices; further demand-driven price increases for lower-GHG fuels such as natural gas; and the costs of equipment and process changes to abate emissions or reduce energy use.

In gauging the potential impacts of GHG regulation, it is important to distinguish the "competitiveness" effect from the broader economic impact on a given industry or firm. A mandatory climate policy will present costs for U.S. firms regardless of what action is taken by other countries. In the case of energy-intensive industries, one potential impact of pricing carbon could be a decline in demand for their products as consumers substitute less GHG-intensive products. This is distinct, however, from the international "competitiveness" impact of GHG regulation, which is only that portion of the total impact on a firm resulting from an imbalance between stronger GHG constraints within, and weaker GHG constraints outside, the United States.

To empirically quantify the potential magnitude of this competitiveness impact, the Pew Center commissioned an analysis by economists at Resources for the Future.[5] This work, which we published in May, analyzes 20 years of data in order to discern the historical relationship between electricity prices and production, consumption, and employment in more than 400 U.S. manufacturing industries. On that basis, the analysis then projects the potential competitiveness impacts of a U.S. carbon price, assuming no comparable action in other countries. The analysis assumes a CO_2 price of \$15 per ton. (EPA's preliminary analysis of the American Clean Energy and Security Act (ACES) Act estimates an allowance price of \$16 per ton CO_2 in 2020.[6])

The Pew/RFF analysis finds an average production decline of 1.3 percent across the U.S. manufacturing sector as a whole, but also a 0.6 percent decline in consumption. This suggests that the decline in production that can be attributed to increased imports – in other words, the competitiveness effect -- is just 0.7 percent. For energy-intensive industries (those whose energy costs exceed 10 percent of shipment value), the analysis projects that average U.S. output declines about 4 percent. However, consumption declines 3 percent, so that only a 1 percent decline in production (or one-fourth of the total decline) can be attributed to an increase in imports, or a loss of competitiveness. For specific energy-intensive industries, including chemicals, paper, iron and steel, aluminum, cement, and bulk glass, the analysis projects a competitiveness impact ranging from 0.6 percent to 0.9 percent, although within certain subsectors, the impact could be higher.

What this analysis demonstrates very clearly is that most of the projected decline in production stems from a reduction in domestic demand for these products, not an increase in imports. In other words, most of the projected economic impact on energy-intensive industries reflects a move toward less emissions-intensive products—as would be expected from any effective climate change policy, even one with global participation—and not a movement of jobs and production overseas. At the carbon price level studied, the projected competitiveness impacts, as well as the broader economic effects on energy-intensive industries, are modest and, in our view, can be readily managed with a range of policy instruments.

POLICY OPTIONS

In the design of a domestic cap-and-trade system, competitiveness concerns can be addressed in part through a variety of cost-containment measures, such as banking and borrowing and the use of offsets, which can help reduce the costs to all firms, including energy intensive, trade-exposed industries. However, other transitional policies may be needed to directly address competitiveness concerns in the period preceding the establishment of an effective international framework. Options include: fully or partially exempting potentially vulnerable firms from the cap-and trade system; compensating firms for the costs of GHG regulation through allowance allocation or tax rebates; transition assistance to help firms adopt lower-GHG technologies, and to help communities and workers adjust to changing labor markets; and border measures such as taxes on energy-intensive imports from countries without GHG controls. In addition, a domestic policy could be designed to encourage and anticipate international sectoral agreements establishing the respective obligations of major producing companies within given sectors.

Exclusion from Coverage

One option is to fully or partially exclude vulnerable sectors or industries from coverage under the cap-and-trade program. For instance, under the Lieberman- Warner Climate Security Act of 2008,[7] the direct "process" emissions of many energy-intensive industries would not have been subject to GHG limits. This type of exclusion would have reduced the number of emission allowances a trade-exposed firm would need to hold and would thereby eliminate some of the direct regulatory costs, shielding it not only from competitiveness impacts but also from some of the broader economic effects of pricing carbon. However, by limiting the scope of the cap-and-trade system, exclusions of trade-exposed industries would undermine the goal of reducing GHG emissions economy-wide, and would reduce the economic efficiency of a national GHG reduction program. Exemptions could also give exempted industries an economic advantage over nonexempt domestic firms and sectors, including competitors. Moreover, firms whose emissions are exempted would still face the indirect costs of higher energy prices and would not be completely shielded from the competitive impact associated with this cost increase.

Compensation for the Costs of GHG Regulation

Another option is to include these sectors in the cap-and-trade system but compensate them for the costs of GHG regulation. Key design considerations include the scope, form, and means of calculating such compensation, and whether and how it should be phased out. As noted earlier, firms covered by the cap-and-trade system face both direct and indirect costs of regulation. Direct compliance costs include the cost of purchasing any allowances needed to cover direct emissions regulated under the cap and/or the cost of equipment and process changes to abate emissions. Indirect costs include higher prices for electricity and natural gas (reflecting an embedded carbon price and, in the case of natural gas, rising demand for this less GHG-intensive fuel). For energy-intensive industries, the indirect cost of higher energy prices represents a significant portion of the total potential cost.

One form of compensation is providing free emission allowances. Because free allocation provides the same economic incentive to reduce emissions as does an auction,[8] keeping energy- intensive sectors under the cap, but providing free allowances, provides for greater environmental effectiveness and economic efficiency than excluding them. Furthermore, additional allowances could be provided to compensate for indirect costs, thus providing a more complete shield from international competitiveness impacts.

Another form of compensation for direct and/or indirect costs could be tax credits or rebates. One potential source of revenue for such measures is proceeds from the auction of emission allowances. A tax rebate would be a direct payment to compensate a firm for GHG regulatory costs; a tax credit could alternatively offset those costs by reducing a non-GHG burden such as corporate or payroll taxes, or healthcare or retirement costs.[9]

Whatever form the compensation takes, the central challenge is determining the appropriate level. In the case of direct compliance costs, allowances could be granted on the basis of historical emissions ("grandfathering") and energy-intensive sectors could receive more generous allocations than other emitters. For instance, energy-intensive industries could receive a full free allocation while others receive 80 percent of their historical emissions. Over time, the energy-intensive sectors could continue to receive a higher proportion of free allowances than other sectors as the allocation system transitions to fuller auctioning. However, granting allowances on the basis of historical emissions can effectively penalize early action and reward relatively heavier emitters within an industry. In addition, it does not necessarily guard against emissions leakage or a loss of jobs, as a firm could choose to maximize profits by selling its free allowances and reducing production. There is also the risk that firms will be over-compensated and realize windfall profits.

Alternatively, compensation could be "output-based," pegged to actual production levels and/or energy consumption. This would shield energy-intensive firms from regulatory costs, and lower the risk of emissions leakage and competitiveness impacts, while providing an incentive for continued production. Firms could be compensated in full for their direct and indirect costs. Or, an output-based approach could incorporate a performance standard (i.e., emissions or energy use per unit of production) to encourage and reward lower GHG-intensity production. For instance, free allowances could be pegged to the level needed by a firm whose emissions intensity is only 85 percent of the sector average; that percentage could decline over time, providing an ongoing incentive to switch to lower-GHG processes and energy sources. This was the approach adopted in the Inslee-Doyle Carbon Leakage Prevention Act introduced in the 110th Congress.[10] The ACES Act adopts an output-based approach, initially allocating 15 percent of the total allowance pool to energy-intensive industries to compensate for both direct and indirect costs based on a facility's level of output. However, as allocations to individual firms would be based on average emissions intensity within the sector, rather than a stronger benchmark, there is no added incentive to improve GHG performance beyond the average.

If compensation is provided, one important consideration is how long it should be maintained and at what level. Phasing out the compensation would give firms additional incentive to improve their GHG performance but would also make them more vulnerable to competitiveness impacts. A mandatory program could provide for periodic review of any allowances or other compensation to vulnerable sectors to consider adjusting them on the basis of new information. For instance, if the legislation establishes a specific timetable for moving from free allocation to auctioning, this transition might be slowed for specific industries if there are clear indications of competitiveness impacts. Alternatively, compensation could be phased out or ended if other countries take stronger action or new international agreements are reached. The ACES Act incorporates such approaches. It would phase down the output-based allowance rebates 10 percent a year starting in 2026, but allow the President to adjust that rate depending on an assessment of emissions leakage.

As with the exclusion of trade-exposed sectors from the cap, the remedy provided by these compensation approaches extends beyond any actual competitiveness effect. Whether based on output or historical emissions, most of the proposals offered to date aim to compensate firms for most or all of the increased costs associated with GHG regulation, not just for the impacts they may face due to the asymmetry between GHG constraints within and outside the United States. To limit compensation to competitiveness impacts alone would require in-depth financial knowledge of each firm and/or complex calculations that could be reliably performed only once the impacts have occurred. A drawback of a broader compensation approach is that the financial resources required—whether drawn from auction revenue or other sources—are not available for other climate- or non-climate-related purposes.

Transition Assistance

Another option is to provide transition assistance to vulnerable firms to help them adopt lower-GHG technologies, and to communities and workers affected by competitiveness impacts. In the case of firms, measures could include tax incentives such as accelerated depreciation to encourage the replacement of inefficient technologies, or tax credits for the development or adoption of lower-GHG alternatives. Firms could also be incentivized to switch to low-carbon energy sources, for example through subsidies for the purchase or generation of renewable energy.

Where competitiveness impacts are unavoidable, assistance can be provided to both workers and communities. Previous government efforts to help communities adjust to economic changes resulting from national policies provide lessons for shaping similar efforts as part of climate change policy.[11] At the level of individual workers, policies such as the Workforce Investment Act providing income support and retraining to help move workers into new jobs can provide a blueprint for transition programs to assist workers adversely affected by competitiveness imbalances under a climate policy.[12] The ACES Act would provide worker transition assistance through two set-asides of emission allowances: one to support retraining and other benefits when employers, unions or other groups of workers demonstrate that employment has suffered as a result of the bill; the other to support training for new jobs in clean energy industries.

Border Adjustment Measures

Another strategy is to try to equalize GHG-related costs for U.S. and foreign producers by imposing a cost or other requirement on energy-intensive imports from countries with weaker or no GHG constraints. One option is a border tax based on an import's "embedded" emissions (equal to the compliance costs for a domestic producer of an equivalent good). Alternatively, under a cap-and-trade system, emission allowances could be required for the import of energy-intensive goods. In the 110th Congress, the Lieberman-Warner bill, the Bingaman-Specter bill, the Markey ICAP bill, and the Dingell-Boucher discussion draft all adopted variations of this approach. Under the ACES Act, "international reserve allowances" would be required for energy-intensive imports starting in 2020 unless a new international agreement meeting the bill's negotiating objectives has entered into force, or unless Congress concurs with the President's determination that the requirement is not in the national interest.

One major shortcoming of unilateral border measures is their limited effectiveness in reducing competitiveness impacts. As the border adjustment measures would apply only to imports to the United States, they would not help "level the playing field" in the larger global market where U.S. producers may face greater competition from foreign producers.

Among the other issues raised by unilateral border measures is their consistency with World Trade Organization (WTO) rules. The legality of a given measure would depend in part on its specific design and on the types of climate policies in place domestically. As such approaches have not been previously employed, there are no definitive rulings, and experts differ in their interpretation of relevant WTO precedents.[13] The legal uncertainties ultimately would be resolved only through the adjudication of a WTO challenge, a likely prospect if unilateral border measures were to be applied by the United States or another country.

Another important consideration is the potential impact on trade and international relations. If the United States were to impose border requirements, there is a greater likelihood that it would become the target of similar measures. European policymakers also are weighing the use of border measures and have argued that the emission targets under consideration in the United States are not comparable to those adopted by the European Union. U.S. trade officials and others also have voiced strong concern about the potential for retaliatory trade measures by targeted countries, leading to escalating trade conflicts.[14] Proponents argue that the threat of unilateral trade measures would give the United States greater leverage in international climate negotiations. However, there is a significant risk that they would engender more conflict than cooperation, in the end making it more difficult to reach agreements that could more effectively address competitiveness concerns.

International Sectoral Agreements

All of the preceding options are measures that would be implemented domestically. Another approach that would help reduce emissions within and outside the United States, while addressing competitiveness concerns, is to negotiate international agreements setting GHG standards or other measures within energy-intensive globally-traded sectors. For example, major steel-producing countries could agree on standards limiting GHGs per ton of steel, which could be differentiated initially according to national circumstances and converge over time. Sectoral agreements could take a number of forms, depending on the specific sectors, and could be stand-alone agreements or integrated into a comprehensive climate framework.[15]

Within the domestic context, a purely sector-by-sector approach would sacrifice the broad coverage and economic efficiency of an economy-wide cap-and-trade program. However, sectoral agreements could exist alongside a cap-and-trade program, and the system could be designed to encourage U.S. producers to work toward their establishment. One option would be to provide for a sector's exclusion from the cap once an international agreement of comparable stringency is in place (although, as noted, diminishing the scope of the cap-and-trade system by exempting one or more sectors would limit its economic efficiency). An alternative is to keep the sectors under the cap but align their obligations under the domestic program and the international sectoral agreement. For instance, a firm's emissions allowance allocation under the trading system could be based on the GHG standard that is agreed to internationally.

In keeping with the principle of "common but differentiated responsibilities," an international sectoral agreement may not set fully equivalent requirements for all countries, particularly at the outset. In that event, compensation for energy-intensive industries could be maintained at some level and phased out as the requirements for other countries rise to those borne by the United States.

RECOMMENDATIONS

Based on our assessment of the available options, the Pew Center believes that the Senate should seek to address competitiveness concerns by:

1. strongly encouraging the executive branch to negotiate a new multilateral climate agreement establishing strong, equitable, and verifiable commitments by all major economies;
2. including in domestic legislation incentives for such an agreement, including support for stronger action by major developing countries; and
3. including in cap-and-trade legislation transitional measures to cushion the impact of mandatory GHG limits on energy-intensive trade-exposed industries and the workers and communities they support. These transitional measures should be structured as follows:
    - In the initial phase of a cap-and-trade program, free allowances should be granted to vulnerable industries to compensate them for the costs of GHG regulation. For direct costs, allowance allocations should be based on actual production levels. For indirect costs, allocations should reflect the emitter's production-based energy consumption, taking into account the GHG intensity of its energy supplies.
    - Based on an analysis of GHG performance within a given sector, allocations should be set initially so that producers with average GHG performance are fully compensated for regulatory costs, while those performing above or below the norm receive allowances whose value is greater or less than their costs,

respectively. This factor should be adjusted over time as an incentive to producers to continually improve their GHG performance.

- Free allocation levels should decline over time, gradually transitioning to full auctioning, although at a slower rate than for other sectors.
- A review should be conducted periodically to assess whether sectors are experiencing competitiveness impacts and, if warranted, to adjust allocation levels and/or the rate of transition to full auctioning.
- A portion of allowance auction revenue should be earmarked for programs to assist workers and communities in cases where GHG constraints are demonstrated to have caused dislocation.
- Transition assistance should be curtailed for a given sector upon entry into force of a multilateral or sectoral agreement establishing reasonable obligations for foreign producers, or upon a Presidential determination that such measures have been instituted domestically.

We believe this approach addresses the transitional competitiveness concerns likely to arise under a mandatory cap-and-trade program, while maintaining the environmental integrity of theprogram and providing an ongoing incentive for producers to improve their GHG performance. We commend the Committee for focusing the attention of the Senate on this critical issue, and would be happy to work with you as you develop legislation to address this and other dimensions of the climate challenge.

I thank you for your attention and would be happy to answer your questions.

End Notes

[1] For more information on the Pew Center on Global Climate Change, please visit http://www.pewclimate.org

[2] For a discussion of how best to encourage strong climate action by other countries, see the testimony on The Roadmap from Poznan to Copenhagen – Preconditions for Success by Elliot Diringer, Vice President for International Strategies for the Pew Center on Global Climate Change, submitted to the Select Committee on Energy Independence and Global Warming, U.S. House of Representatives, February 4, 2009. (http://www.pewclimate.org/testimony/diringer/02-04-09)

[3] EPA *Analysis of the American Clean Energy and Security Act of 2009* H.R. 2454 in the 111th Congress 6/23/09: Data Annex http://www.epa.gov/climatechange/economics/economicanalyses.html

[4] Jaffe, A.B., S.R. Peterson, P.R. Portney, and R.N. Stavins, "Environmental Regulation and the Competitiveness of U.S. Manufacturing: What Does the Evidence Tell Us?," Journal of Economic Literature, Vol. 23, March 1995.

[5] Aldy, J.E. and Pizer, W. A., *The Competitiveness Impacts of Climate Change Mitigation Policies*, Pew Center on Global Climate Change, May 2009. http://www.pewclimate.org/international/CompetitivenessImpacts.

[6] EPA Analysis of the American Clean Energy and Security Act of 2009 H.R. 2454 in the 111th Congress 6/23/09http://www.epa.gov/climatechange/economics/pdfs/HR2454_Analysis.pdf .

[7] S.3036 of the 110th Congress.

[8] The cap in a "cap-and-trade" system determines its environmental stringency by setting the number of emission allowances that are available. These allowances are equal to the amount of emissions that are permitted under the cap and their number declines over time as the cap is tightened. From an environmental perspective, it doesn't matter how the emission allowances are distributed. They could be auctioned or freely distributed or any combination of the two. All that matters is the total number of emission allowances that are distributed -- the environmental goal is determined by the cap itself and is not in any way impacted by whether the allowances are auctioned or distributed freely. A company that is included in the cap-and-trade program but given free allowances still has an incentive to reduce its emissions because that would free up allowances that the company could sell.

[9] Houser, Trevor et al., *Leveling the Carbon Playing Field: International Competition and US Climate Policy Design*, Peterson Institute for International Economics and World Resources Institute, May 2008.

[10] H.R. 7146 in the 110th Congress.

[11] Greenwald, Judith M., Brandon Roberts, and Andrew D. Reamer, *Community Adjustment to Climate Change Policy*, Pew Center on Global Climate Change, December 2001.

[12] Barrett, Jim, *Worker Transition and Global Climate Change*, Pew Center on Global Climate Change, December 2001.

[13] For a discussion of WTO-related issues, see Bordoff, Jason E., *International Trade Law and the Economics of Climate Policy: Evaluating the Legality and Effectiveness of Proposals to Address Competitiveness and Leakage Concerns*, Brookings Institution, June 2008.

[14] Remarks of U.S. Trade Representative Susan C. Schwab to U.S. Chamber of Commerce, January 17, 2008.

[15] Bodansky, Daniel, *International Sectoral Agreements in a Post-2012 Climate Framework*, Pew Center on Global Climate Change, May 2007.

In: Climate Change Measures and Trade Considerations ISBN: 978-1-60876-756-4
Editor: George R. Fried

Chapter 3

CLIMATE CHANGE LEGISLATION: INTERNATIONAL TRADE CONSIDERATIONS

Chuck Grassley

Today's hearing will explore the international trade implications of possible carbon emissions legislation, also known as "cap-and-trade" legislation. The House has passed a bill, and the Senate is about to begin its legislative process as well. So, it's important that the Finance Committee explore these issues, because Congress could be setting our manufacturers up for a lot of trouble if we don't. I've said many times that we ought to approach this issue through a worldwide, international agreement. That's the only way to ensure that China and India and other major carbon emitting countries are involved. Otherwise, our industry is going to be left very uncompetitive. We're going to see more manufacturing move overseas where less efficient plants produce far more pollution than our American industries, and nobody should want to do that.

Some in Congress think the answer to that problem is to include some type of border measure in any legislation that Congress passes. At least that's the view of the House of Representatives. But I'm skeptical of that approach. I think it would be difficult to design such an approach that would be consistent with the rules of the World Trade Organization. If we do something that's inconsistent with the rules of the World Trade Organization, we're opening ourselves up to trade sanctions that could dwarf any sanctions we've seen in the past.

And then we'll have hurt our manufacturers twice. We'll have raised their costs by imposing cap and trade in the first place. And then we'll compound the problem by giving foreign countries a license to hit us with sanctions. That wouldn't be good for us, and it wouldn't be good for the international trading system either.

And it could get even worse. Other countries could follow our lead and impose their own border measures against our exporters. We could find ourselves defending our measures at the World Trade Organization and challenging other countries' measures at the same time.

I understand that the World Trade Organization would prefer to avoid that scenario and not have this issue thrust into its lap. The Director-General of the World Trade Organization, Pascal Lamy, stated recently that the WTO membership does not want to decide what is or isn't allowed from a trade perspective. He indicated that the World Trade Organization would much rather have the trade issues addressed as part of whatever comes out of the meetings scheduled to take place in Copenhagen later this year.

That makes sense. If the United States unilaterally imposes a border measure, we'll make it that much harder to reach an international agreement. Other countries aren't going to want to negotiate with us if they think we're dictating a specific outcome. Border measures aren't the only approach that has been suggested for addressing the competitiveness issue.

Another suggestion is to give free emissions allowances to domestic industries that are deemed to be trade sensitive. That's the approach that the European Union took. I have some questions about that approach too. I'm not convinced that giving away free allowances would be consistent with the subsidy rules of the World Trade Organization. At a minimum, there would be a risk of other countries arguing that free emission permits are subsidies that cause "adverse effects" to their industries. If we lost a subsidy case, we'd have the same risk of trade sanctions that I mentioned earlier. So for this reason too, we'd be better off waiting for an international agreement instead of pushing ahead unilaterally. I look forward to exploring these issues in further detail with our witnesses, and I thank each of them for appearing here today.

Chapter 4

SUMMARY TESTIMONY OF GARY N. HORLICK BEFORE THE UNITED STATES SENATE COMMITTEE ON FINANCE

Gary N. Horlick

Thank you, Mr. Chairman, and Senator Gras sley, for the opportunity to appear before this Committee. I wish to state at the outset that I am not appearing on behalf of any client or any other organization or person.

The issues presented by climate change are extremely complex, and seem to evoke extremely complex solutions. I wish to start, however, with two basic rules to apply to these very complicated questions.

1. Before we adopt any measures in our legislation, would we like it if other countries adopted the same measures? There are good practical reasons for this old-fashioned rule. The U.S. exports about 20% of its manufactured goods. And 30% of U.S. manufactured exports are by small and medium companies. We export about 30% of our agricultural production, which is the difference between a profit and a loss for many of America's farmers.
2. Is the measure consistent with the obligations we have undertaken, most obviously through the WTO Agreements, but also our other agreements? There are also very practical reasons for following this rule. It is tempting to say that we can re-interpret existing WTO rules to permit whatever measures are necessary to protect our environment. But do we really want to change those existing rules? The key to the U.S. economy is constant innovation. One of the important fields where we lead the world of innovation is biotechnology, which is revolutionizing medicine, agriculture, and even many of the environmental concerns dealt with in proposed legislation (such as environmental remediation and renewable fuels). So far the United States has resisted efforts in Europe and elsewhere to limit our market access for our products because of how they are produced – from biotech means. But if we re-interpret WTO rules to allow trade barriers based on how things are made, we open

up a can of worms – and might permit other countries to block our biotech exports, including major items such as corn, soybeans, and other crops.

With these two points in mind, I would like to address some of the international trade issues that have arisen in the context of the climate change debate:

1. The WTO itself, in a recent report with UNEP, has pointed out that some *border measures* would be permissible as part of a climate change package. But the WTO very carefully stated that WTO rules "permit, *under certain conditions"* (emphasis added) the use of border measures and indeed the devil is in the details. A VAT-style tax, imposed identically on domestic and imported goods, should past muster, but after that, it gets very difficult to design a border tax which would pass muster, as we have seen with prior GATT and WTO litigation. Requiring emission permits for importers – especially if there are any restrictions on those permits, or if any are given away to domestic producers – would raise serious WTO questions. But before we even get to those legal issues, it may be worth pondering the practical ones.

   As noted above, do we want similar import taxes or permit requirements imposed on our exports. Do we want each country finding a separate basis for imposing border restrictions? Do we want countries with stricter emissions standards than our factory emissions standards blocking our exports? As President Obama pointed out, India and China have lower per capita emissions than we do. Do we want developing countries, including not only China and India, but indeed most countries in the world, blocking our exports on the grounds that our per capita greenhouse gas emissions are greater than theirs?

   Border measures would raise costs for some U.S. manufacturers – just as they have to adapt to new climate change standards.

   In practice, it seems that import restrictions are much more likely to be challenged in the WTO than is financial assistance to producers, such as offsetting costs or giving away permits. Thousands of pages of non-export subsidies are reported to the WTO, but only a handful have been challenged in the WTO, while literally hundreds of border measures have been challenged in the WTO and its predecessor GATT.
2. *Permit allocation* is not prohibited by the WTO rules – but, again, it depends on the details of the design. Linking permits to export requirements, or requirements to use domestic goods, would be prohibited. And even without those problems, permit allocations could be challenged in the WTO as an "actionable" subsidy. WTO rules require objective criteria -- rules insisted on by the United States. The same subsidy rules would also be relevant in designing incentives for developing new sources of energy, and even renewable energy standards.
3. Obviously, no one country, even one as large as the United States, can clean up Earth's atmosphere on its own. It is essential that all major emitters co-operate in this

effort. U.S. efforts will fail if other emitters do not do their part. I have great respect for the U.S. negotiator, Todd Stern, and his team. They certainly have a difficult job, and one which will require close cooperation between Congress and the executive as the U.S. pursues an international agreement. Fortunately, all the major countries realize that they have a stake in this game -- some major developing countries face rapidly increasing desertification if GHG emissions are not reduced worldwide, including by those same countries. Even though there are shared interests, the climate change negotiation will not be easy, and numerous tactical choices will have to be made, such as whether to complete the legislation before the Copenhagen talks. If we make choices which reduce the chances of a good international agreement, we harm ourselves as well.

4. Perhaps the biggest international trade challenge -- and one on which a lot more work needs to be done -- is how the mechanics of international trade will work if each of the hundred and ninety countries (or even 10-15 regional groupings) has its own individual climate change implementation. What if some of them have border taxes, some require permits for imports, and others instead offset the costs for their domestic industry. Or each country has a capand-trade system with different limitations on the permits? These are highly practical questions, but, in the end, they will determine the true economic impact of the climate change legislation and agreements.

According to a recent study, an Apple iPOD with a retail price of $299 arrives in the U.S. in a box labeled Made In China, but only $4 stays in China while $130 stays in the U.S. and the remainder is scattered among 18 other countries. How are the 20 countries that collectively make an Apple iPOD going to calculate all the carbon impacts and untangle all those permits, taxes, and rebates? These details need to be worked out carefully – and in advance.

Thank you very much, I would be happy to answer any questions.

In: Climate Change Measures and Trade Considerations ISBN: 978-1-60876-756-4
Editor: George R. Fried

Chapter 5

CLIMATE CHANGE TRADE MEASURES: ESTIMATING INDUSTRY EFFECTS

Loren Yager

Mr. Chairman and Members of the Committee:

Thank you for the opportunity to appear again before the Committee to provide insights from GAO's work on issues related to important international issues. Changes in the earth's climate attributable to increased concentrations of greenhouse gases may have significant environmental and economic impacts in the United States and internationally. To mitigate climate change effects, countries are taking or considering varying approaches to reducing greenhouse gas emissions, such as carbon dioxide, which is the most important greenhouse gas due to its significant volume. Between 2007 and 2009, Congress introduced a number of climate change bills, many of which contained proposals for a domestic emissions pricing system, such as a cap-and-trade system or a carbon tax. However, imposing costs on energy-intensive industries in the United States could potentially place them at a disadvantage to foreign competitors. In addition, emissions pricing could have negative environmental consequences, such as "carbon leakage," whereby emissions reductions in the United States are replaced by increases in production and emissions in less-regulated countries. As Congress considers the design of a domestic emissions pricing system, a key challenge will be balancing the need to reduce greenhouse gas emissions with the need to address the competitiveness of U.S. industries.

In my testimony today, my comments are based on a report that we are issuing today to the Senate Committee on Finance.[1] In particular, I will briefly describe some of the key challenges associated with estimating the industry effects from climate change measures, and provide illustrations of key characteristics for potentially vulnerable industries.

To address these objectives, we interviewed officials and reviewed climate change literature and documents from U.S. agencies, international organizations, policy institutes, and professional organizations; reviewed and analyzed a selection of climate change legislation introduced between 2007 and 2009 and congressional hearing records; analyzed data from the Census Bureau and the Departments of Energy and Commerce, among others; and reviewed and presented summary results for studies attempting to quantify the potential international competitiveness effects on domestic industries from emissions pricing. We conducted our work from October 2008 to July 2009 in accordance with all sections of GAO's Quality Assurance Framework that were relevant to our objectives. We believe that the information and data obtained, and the analysis conducted, provide a reasonable basis for any findings and conclusions in this product.

SUMMARY

Estimating the potential effects of domestic emissions pricing for industries in the United States is complex. If the United States were to regulate greenhouse gas emissions, production costs could rise for certain industries and could cause output, profits, or employment to fall. Within these industries, some of these adverse effects could arise through an increase in imports, a decrease in exports, or both. However, the magnitude of these potential effects is likely to depend on the greenhouse gas intensity of industry output and on the domestic emissions price, which is not yet known, among other factors.

Estimates of adverse competitiveness effects are generally larger for industries that are both relatively energy- and trade-intensive. In 2007, these industries accounted for about 4.5 percent of domestic output. Estimates of the effects vary because of key assumptions required by economic models. For example, models generally assume a price for U.S. carbon emissions, but do not assume a similar price by other nations. In addition, the models generally do not incorporate all policy provisions, such as legislative proposals related to trade measures and rebates that are based on levels of production.

Proposed legislation suggests that industries vulnerable to competitiveness effects should be considered differently. Industries for which competitiveness measures would apply are identified on the basis of their energy and trade intensity. Most of the industries that meet these criteria are in primary metals, nonmetallic minerals, paper, and chemicals, although significant variation exists for product groups (sub-industries) within each industry. Additional variation arises on the basis of the type of energy used and the extent to which foreign competitors' greenhouse gas emissions are regulated. To illustrate variability in characteristics that make industries vulnerable to competitiveness effects, we include illustrations of sub-industries within primary metals that meet both the energy and trade intensity criteria; examples that met only one criterion; and examples that met neither, but had significant imports from countries without greenhouse gas pricing.

BACKGROUND

Countries can take varying approaches to reducing greenhouse gas emissions. Since energy use is a significant source of greenhouse gas emissions, policies designed to increase energy efficiency or induce a switch to less greenhouse-gas-intensive fuels, such as from coal to natural gas, can reduce emissions in the short term. In the long term, however, major technology changes will be needed to establish a less carbon- intensive energy infrastructure. To that end, a U.S. policy to mitigate climate change may require facilities to achieve specified reductions or employ a market-based mechanism, such as establishing a price on emissions. Several bills to implement emissions pricing in the United States have been introduced in the 110th and 111th Congresses. These bills have included both cap-and-trade and carbon tax proposals. Some of the proposed legislation also include measures intended to limit potentially adverse impacts on the international competitiveness of domestic firms.

ESTIMATING COMPETITIVENESS EFFECTS

Estimating the effects of domestic emissions pricing for industries in the United States is complex. For example, if the United States were to regulate greenhouse gas emissions, production costs could rise for many industries and could cause output, profits, or employment to fall. However, the magnitude of these potential effects is likely to depend on the greenhouse gas intensity of industry output and on the domestic emissions price, which is not yet known, among other factors. Additionally, if U.S. climate policy was more stringent than in other countries, some domestic industries could experience a loss in international competitiveness. Within these industries, adverse competitiveness effects could arise through an increase in imports, a decrease in exports, or both.

For regulated sources, greenhouse gas emissions pricing would increase the cost of releasing greenhouse gases. As a result, it would encourage some of these sources to reduce their emissions, compared with businessas-usual. Under domestic emissions pricing, production costs for regulated sources could rise as they either take action to reduce their emissions or pay for the greenhouse gases they release. Cost increases are likely to be larger for production that is relatively greenhouse gas-intensive, where greenhouse gas intensity refers to emissions per unit of output. Cost increases may reduce industry profits, or they may be passed on to consumers in the form of higher prices. To the extent that cost increases are passed on to consumers, they could demand fewer goods, and industry output could fall.

While emissions pricing would likely cause production costs to rise for certain industries, the extent of this rise and the resulting impact on industry output are less certain due to a number of factors. For example, the U.S. emissions price and the emissions price in other countries are key variables that will help to determine the impact of emissions pricing on domestic industries. However, future emission prices are currently unknown. Additionally, to the extent that emissions pricing encourages technological change that reduces greenhouse gas intensity, potential adverse effects of emissions pricing on profits or output could be mitigated for U.S. industries.

Several studies by U.S. agencies and experts have used models of the economy to simulate the effects of emissions pricing policy on output and related economic outcomes. These models generally find that emissions pricing will cause output, profits, or employment to decline in sectors that are described as energy intensive, compared with business-as-usual. In general, these studies conclude that these declines are likely to be greater for these industries, as compared with other sectors in the economy. However, some research suggests that not every industry is likely to suffer adverse effects from emissions pricing. For example, a long-run model estimated by Ho, Morgenstern, and Shih (2008) predicts that some U.S. sectors, such as services, may experience growth in the long run as a result of domestic emissions pricing.[2] This growth would likely be due to changes in consumption patterns in favor of goods and services that are relatively less greenhouse gas-intensive.

Potential international competitiveness effects depend in part on the stringency of U.S. climate policy relative to other countries. For example, if domestic greenhouse gas emissions pricing were to make emissions more expensive in the United States than in other countries, production costs for domestic industries would likely increase relative to their international competitors, potentially disadvantaging industries in the United States. As a result, some domestic production could shift abroad, through changes in consumption or investment patterns, to countries where greenhouse gas emissions are less stringently regulated. For example, consumers may substitute some goods made in other countries for some goods made domestically. Similarly, investment patterns could shift more strongly in favor of new capacity in countries where greenhouse gas emissions are regulated less stringently than in the United States.

Stakeholders and experts have identified two criteria, among others, that are important in determining potential vulnerability to adverse competitiveness effects: trade intensity and energy intensity. Trade intensity is important because international competitiveness effects arise from changes in trade patterns. For example, if climate policy in the United States were more stringent than in other countries, international competition could limit the ability of domestic firms to pass increases in costs through to consumers. Energy intensity is important because the combustion of fossil fuels for energy is a significant source of greenhouse gas emissions, which may increase production costs under emissions pricing.

Legislation passed in June 2009 by the House of Representatives, H.R. 2454, 111th Cong. (2009), uses the criteria of trade intensity and energy intensity or greenhouse gas intensity, among others, to determine eligibility for the Emission Allowance Rebate Program, which is part of the legislation.[3] H.R. 2454 specifies how to calculate the two criteria. Trade intensity is defined as the ratio of the sum of the value of imports and exports within an industry to the sum of the value of shipments[4] and imports within the industry. Energy intensity is defined as the industry's cost of purchased electricity and fuel costs, or energy expenditures, divided by the value of shipments of the industry.

Reducing carbon emissions in the United States could result in carbon leakage through two potential mechanisms. First, if domestic production were to shift abroad to countries where greenhouse gas emissions are not regulated, emissions in these countries could grow faster than expected otherwise. Through this mechanism, some of the expected benefits of reducing emissions domestically could be offset by faster growth in emissions elsewhere, according to Aldy and Pizer (2009).[5]

Second, carbon leakage may also arise from changes in world prices that are brought about by domestic emissions pricing. For example, U.S. emissions pricing could cause domestic demand for oil to fall. Because the United States is a relatively large consumer of oil worldwide, the world price of oil could fall when the U.S. demand for oil drops. The quantity of oil consumed by other countries would rise in response, increasing greenhouse gas emissions from the rest of the world. These price effects may be a more important source of carbon leakage than the trade effects previously described.

Potentially Vulnerable Industries

Two key indicators of potential vulnerability to adverse competitiveness effects are an industry's energy intensity and trade intensity. Proposed U.S. legislation specifies that (1) either an energy intensity or greenhouse gas intensity of 5 percent or greater; and (2) a trade intensity of 15 percent or greater be used as criteria to identify industries for which trade measures or rebates would apply. Since data on greenhouse gas intensity are less complete, we focused our analysis on industry energy intensity. Most of the industries that meet these criteria fall under 4 industry categories: primary metals, nonmetallic minerals, paper, and chemicals. However, there is significant variation in specified vulnerability characteristics among different product groups ("sub-industries").

Although our report examined the four industry categories, figures 1 through 4 or the following pages illustrate the variation among different sub-industries within the primary metals industry, as well as information on the type of energy used and location of import and export markets.[6] The data shown in these figures are for the latest year available.

As shown by sub-industry examples in figure 1, energy and trade intensities differ within primary metals. For example, primary aluminum meets the vulnerability criteria with an energy intensity of 24 percent and a trade intensity of 62 percent. Ferrous metal foundries meets the energy intensity criteria, but not the trade intensity criteria. Steel manufacturing—products made from purchased steel—and aluminum products fall short of both vulnerability criteria. Iron and steel mills has an energy intensity of 7 percent and a trade intensity of 35 percent and is by far the largest sub-industry example, with a 2007 value of output of over \$93 billion. The energy and trade intensity for all primary metal products is denoted by the "x" in figure 1.

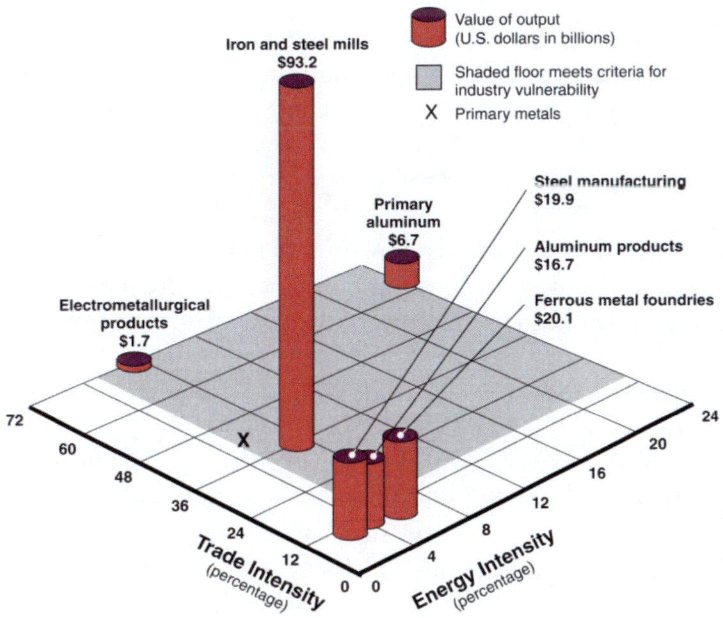

Source: GAO analysis of Department of Commerce energy data for 2006 and trade data for 2007.V.

Figure 1. Energy and Trade Intensity Indicators for Primary Metals Sub-Industry Categories

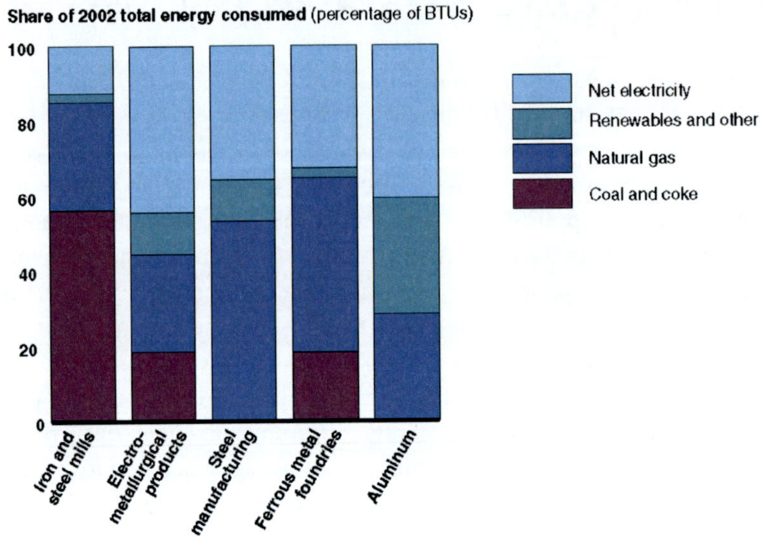

Source: GAO analysis of data from the Department of Energy.

Figure 2. Type of Energy Used by Primary Metals Sub-Industry Categories.

Among the primary metals sub-industry examples shown in figure 2, the types of energy used also vary. Iron and steel mills uses the greatest share of coal and coke, and steel manufacturing and ferrous metal foundries uses the greatest proportion of natural gas. Since coal is more carbon- intensive than natural gas, sub-industries that rely more heavily on coal could also be more vulnerable to competitiveness effects. The carbon intensity of electricity, used heavily in the production of aluminum, will also vary on the basis of the source of energy used to generate it and the market conditions where it is sold. Data shown for "aluminum" include primary aluminum and aluminum products, and net electricity is the sum of net transfers plus purchases and generation minus quantities sold.

Industry vulnerability may further vary depending on the share of trade with countries that do not have carbon pricing. To illustrate this variability, figure 3 provides data on the share of imports by source, since imports exceed exports in each of the primary metals examples. As shown, while primary aluminum is among the most trade-intensive, the majority of imports are from Canada, an Annex I country with agreed emission reduction targets.[7] For iron and steel mills, over one-third of imports are from the European Union and other Annex I countries, not including Canada ("EU plus"). However, for iron and steel mills, almost 30 percent of imports are also from the non-Annex I countries of China, Mexico, and Brazil. While less trade-intensive, steel manufacturing and aluminum products each has greater than one-third of imports from China alone.

As shown in figure 4, adverse competitiveness effects from emissions pricing could increase the already growing share of Chinese imports that exists in some of the sub-industries. Among the examples, iron and steel mills, steel manufacturing, and aluminum products exhibit a growing trade reliance on Chinese imports since 2002. This trend has largely been driven by lower labor and capital costs in China, and, according to representatives from the steel industry, China has recently been producing 50 percent of the world's steel.

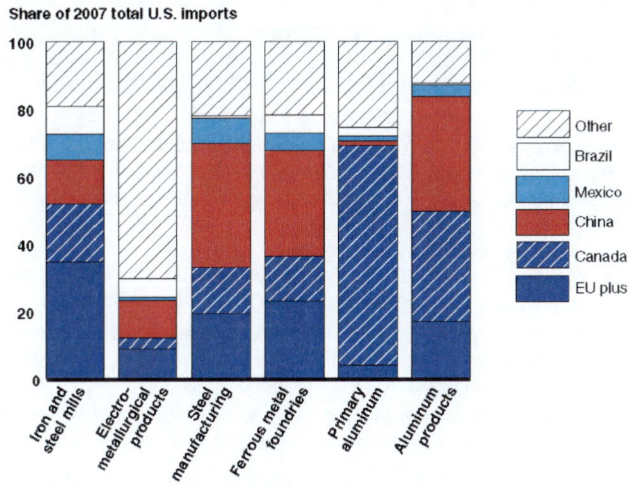

Source: GAO analysis of data from the Department of Commerce.

Figure 3. Source of Imports for Primary Metals Sub-Industry Categories.

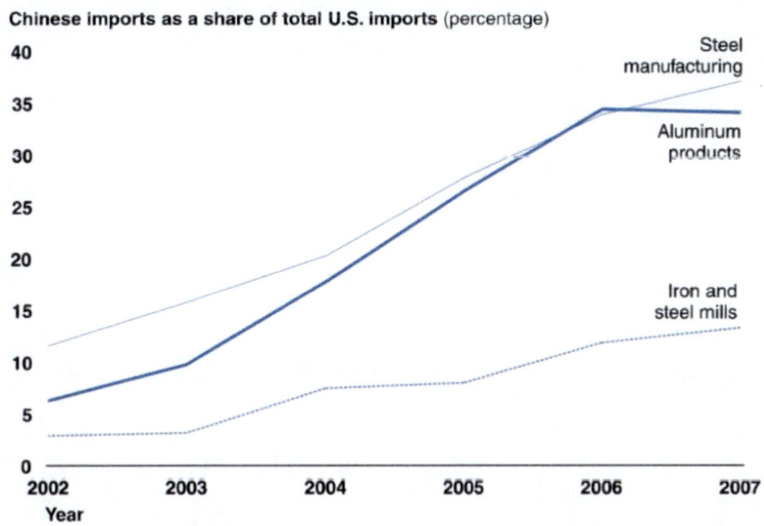

Total U.S. imports by value (U.S. dollars in millions)						
Year	**2002**	**2003**	**2004**	**2005**	**2006**	**2007**
Iron and steel mills	$12,558	10,808	23,355	25,131	33,060	30,445
Steel manufacturing	$1,110	1,184	1,792	1,884	1,916	1,822
Aluminum products	$498	549	718	913	1,209	1,183

Source: GAO analysis of data from the Department of Commerce.

Figure 4. Chinese Imports as Share of Primary Metals Sub-Industry Categories

Mr. Chairman, this concludes my prepared statement. Thank you for the opportunity to testify before the Committee on some of the issues addressed in our report on the subject of climate change trade measures. I would be happy to answer any questions from you or other members of the Committee.

End Notes

[1] GAO, *Climate Change Trade Measures: Considerations for U.S. Policymakers*, GAO-09-724R (Washington, D.C.: July 8, 2009).

[2] Ho, Mun S., Richard Morgenstern, and Jhih-Shyang Shih. (November 2008) "Impact of Carbon Price Policies on U.S. Industry." RFF Discussion Paper No. 08-37, Resources for the Future, Washington, D.C.

[3] Proposed legislation specifies that, in addition to trade intensity, either energy intensity or greenhouse gas intensity should be considered.

[4] An industry's value of shipments represents its value of output.

[5] Aldy, Joseph E. and Pizer, William A. (May 2009) "The Competitiveness Impacts of Climate Change Policies." Pew Center on Global Climate Change, Arlington, VA.

[6] For examples in nonmetallic minerals, paper, and chemicals, as well as further information on data sources and our methodology, see GAO-09-724R.

[7] Annex I countries are parties to the United Nations Framework Convention on Climate Change that are industrialized countries and were members of the Organization for Economic Cooperation and Development in 1992, plus countries characterized as economies in transition.

In: Climate Change Measures and Trade Considerations
Editor: George R. Fried ISBN: 978-1-60876-756-4

Chapter 6

CLIMATE CHANGE TRADE MEASURES: CONSIDERATIONS FOR U.S. POLICY MAKERS

Loren Yager

July 8, 2009
The Honorable Max Baucus
Committee on Finance, Chairman United States Senate

Subject: *Climate Change Trade Measures: Considerations for U.S. Policy Makers*

Dear Mr. Chairman:

Changes in the earth's climate attributable to increased concentrations of greenhouse gases may have significant environmental and economic impacts in the United States and internationally. To mitigate climate change effects, countries are taking or considering varying approaches to reducing greenhouse gas emissions, such as carbon dioxide, which is the most important greenhouse gas due to its significant volume. These approaches range from measures to increase energy efficiency to longer-term efforts to develop technologies to establish a less carbon-intensive energy infrastructure. A U.S. policy to mitigate climate change may require domestic production facilities to achieve specified reductions or employ a market-based mechanism, such as establishing a price on emissions.

Between 2007 and 2009, Congress introduced a number of climate change bills, many of which contained proposals for a domestic emissions pricing system, such as a cap-and-trade system[1] or a carbon tax[2]. Through greenhouse gas emissions pricing, governments create an incentive for parties to lower their greenhouse gas emissions by placing a cost on them. However, imposing costs on energy-intensive industries in the United States could potentially place them at a disadvantage to foreign competitors. In addition, emissions pricing could have negative environmental consequences, such as "carbon leakage," whereby emissions reductions in the United States are replaced by increases in production and emissions in less-regulated countries. As the Congress considers the design of a domestic emissions pricing system, a key challenge will be balancing the need to reduce greenhouse gas emissions with the need to address the competitiveness of U.S. industries.

In anticipation of Senate deliberation on climate change legislative proposals, you asked us to examine how greenhouse gas emissions pricing could potentially affect the international competitiveness of U.S. industries, and to examine trade measures being considered as part of proposed U.S. climate change legislation.[3] On July 6, 2009, we briefed your staff covering: (1) what is known about estimating industry effects, (2) examples of industries that may be vulnerable to a loss in international competitiveness from emissions pricing, (3) trade measures and other approaches to address competitiveness issues, and (4) the potential international implications of trade measures. A copy of the slides presented at the briefing is attached.

To address these objectives, we interviewed officials and reviewed climate change literature and documents from relevant federal agencies, international organizations, policy institutes, businesses, professional organizations, and universities; reviewed and analyzed climate change legislation introduced between 2007 and 2009 and congressional hearing records; we reviewed and presented summary results for two studies that attempt to quantify the potential international competitiveness effects on domestic industries from greenhouse gas emissions pricing; and conducted interviews with officials from the embassies of Australia, Brazil, Canada, China, the European Union, and Mexico. We conducted our work from October 2008 to July 2009 in accordance with all sections of GAO's Quality Assurance Framework that are relevant to our objectives. The framework requires that we plan and perform the engagement to obtain sufficient and appropriate evidence to meet our stated objectives and to discuss any limitations in our work. We believe that the information and data obtained, and the analysis conducted, provide a reasonable basis for any findings and conclusions in this product. For additional details regarding our scope and methodology, see appendix I.

We are sending copies of this report to interested congressional committees and to other parties. In addition, the report will be available at no charge on GAO's Web site at http://www.gao.gov.

If you or your staff have any questions or wish to discuss this material further, please contact me at (202) 512-4347 or yagerl@gao.gov. Contact points for our Offices of Congressional Relations and Public Affairs may be found on the last page of this report. Christine Broderick (Assistant Director), Etana Finkler, Kendall Helm, Jeremy Latimer, Maria Mercado, and Ardith Spence, made significant contributions to this report. In addition, Karen Deans, David Dornisch, Grace Lui, and Jena Sinkfield provided technical assistance.

Sincerely yours,

Loren Yager
Director, International Affairs and Trade

WHY GAO DID THIS STUDY

Global climate change is one of the most significant long-term policy challenges facing the United States, and policies to mitigate climate change will have important economic, social, and environmental implications.

Members of Congress have introduced several bills to address the problem of climate change, many of which establish domestic emissions pricing by requiring firms that emit greenhouse gases either to pay a tax or to hold emission allowances. Whichever approach is taken, domestic emissions pricing could produce environmental benefits by encouraging U.S. firms to reduce their emissions of greenhouse gases. But such pricing could also harm U.S. firms' competitiveness, especially in energy-intensive industries where firms compete internationally. Additionally, there could be increased emissions abroad if production were to increase in other countries as a result of increased domestic costs of production resulting from a U.S. climate policy (carbon leakage). To help reduce impacts on U.S. firms and prevent carbon leakage, several climate change bills have also included trade measures or output-based rebates. The bills have included trade measures that would require importers to purchase emission allowances or pay a border tax for the greenhouse gas emissions associated with their imports. They have also designed output-based rebates to financially rebate industries for the costs incurred under a domestic emissions pricing system.

Objectives, Scope, and Methodology

GAO was asked to examine the potential effects of greenhouse gas emissions pricing on U.S. industries' international competitiveness and trade measures being considered as part of U.S. legislative proposals to address climate change.

Specifically, we examined: (1) what is known about estimating industry effects; (2) examples of industries that may be vulnerable to a loss in international competitiveness from emissions pricing; (3) trade measures and other approaches to address competitiveness issues; and (4) potential international implications of trade measures.

To address these objectives, we interviewed officials and reviewed climate change literature and documents from U.S. agencies, international organizations, policy institutes, and professional organizations; reviewed and analyzed climate change legislation introduced between 2007 and 2009 and congressional hearing records; reviewed and presented summary results for two studies attempting to quantify the potential international competitiveness effects on domestic industries from emissions pricing; and conducted interviews with officials from the embassies of Australia, Brazil, Canada, China, the European Union, and Mexico.

The Office of the United States Trade Representative provided technical comments on this report.

The analysis in this report reflects changes to the American Clean Energy and Security Act of 2009 (HR 2454), as of June 26, 2009.

For additional details regarding our scope and methodology, see appendix I.

Estimating Industry Effects. Estimating the potential effects of domestic emissions pricing for industries in the United States is complex. If the United States were to regulate greenhouse gas emissions, production costs could rise for certain industries and could cause output, profits, or employment to fall. Within these industries, some of these adverse effects could arise through an increase in imports, a decrease in exports, or both. Estimates of adverse competitiveness effects are generally larger for industries that are both relatively energy and trade intensive. In 2007, these industries accounted for about 4.5 percent of domestic output.

Estimates of the effects vary because of key assumptions required by economic models. For example, models generally assume a price for U.S. carbon emissions, but do not assume a similar price by other nations. In addition, the models generally do not incorporate all policy provisions, such as legislative proposals related to trade measures and rebates that are based on levels of production.

Potentially Vulnerable Industries. Proposed legislation suggests that industries vulnerable to competitiveness effects should be considered differently. Industries for which competitiveness measures would apply are identified on the basis of their energy and trade intensity. Most of the industries that meet these criteria are in primary metals, nonmetallic minerals, paper, and chemicals, although significant variation exists for product groups (sub-industries) within each industry. Additional variation arises on the basis of the type of energy used and the extent to which foreign competitors' greenhouse gas emissions are regulated.

To illustrate variability in characteristics that make industries vulnerable to competitiveness effects, we selected example sub-industries within primary metals, non metallic minerals, paper products, and chemicals based on multiple factors. For example, we selected sub-industries that met both the energy and trade intensity criteria, examples that met only one criterion, and examples that met neither, but had significant imports from countries without greenhouse-gas pricing.

Trade Measures. Trade measures have been proposed to help address potential industry and environmental effects of a domestic emissions pricing system; however, questions exist about their proposed effectiveness. Supporters argue that trade measures may help prevent a decline in output by U.S. producers, prevent carbon leakage, and create leverage for other countries to reduce emissions. Opponents raise concerns that trade measures may motivate retaliatory actions, undermine efforts to secure multilateral consensus, and generate little leverage. In addition, potential implementation challenges may exist.

Output-based rebates are measures that offset incurred costs to industries due to emissions pricing and can take the form of a per-unit rebate, allocation of allowances or tax credit. These rebates have been proposed to address competitiveness effects; however, limitations with these measures may also exist. For example, output-based rebates may increase costs for industries that do not receive them, under certain conditions.

How a trade measure is ultimately designed will have important implications for the effectiveness of the overall measure in addressing industry and environmental effects. For example, delaying the timing for when the trade measure goes into effect could result in the measure being a more effective leveraging tool, and would provide other countries an opportunity to implement similar carbon mitigation policies to reduce global emissions. On the other hand, having the trade measure go into effect simultaneously when the domestic cap-and-trade system goes into effect may reduce impacts on U.S. industry competitiveness.

International Trade Implications. A unilateral U.S. trade measure could have important international implications on U.S. bilateral and multilateral trade relations. For example, other countries could view U.S. trade measures as trade restrictions or sanctions, which could lead them to implement restrictions against U.S. exports. Attention also has focused on the potential for trade measures or output-based rebates to be challenged under World Trade Organization (WTO) rules.

Assessing the international trade implications is difficult for a number of reasons. One is that it depends in part upon how other nations reduce their carbon emissions, and whether they perceive any U.S. measures as likely to affect their exports. In addition, the outcome of any WTO challenge, if any, would be uncertain and may depend on how the measure is implemented.

Background

Greenhouse Gas Emissions

The **Intergovernmental Panel on Climate Change (IPCC)** was established to provide scientific and objective information on climate change. According to the IPCC, global atmospheric concentrations of **greenhouse gases** have increased markedly as a result of human activities over the past 200 years, contributing to a warming of the earth's climate. Climate change is a long-term and global issue because greenhouse gases disperse widely in the atmosphere once emitted and can remain there for an extended period of time. Among other potential impacts, climate change could threaten coastal areas with rising sea levels, alter agricultural productivity, and increase the intensity and frequency of floods and tropical storms. While the effect of increases in greenhouse gas emissions on ecosystems and economic growth is expected to vary across regions, countries, and economic sectors, a panel of 18 climate change economists convened by GAO in coordination with the National Academy of Sciences agreed that the Congress should consider using a market-based mechanism to place a price on emissions, and 14 of the 18 panelists were at least moderately confident that the benefits of taking these actions would outweigh the costs. See **GAO-08-605**.

As shown in figure 1, the United States emitted more carbon dioxide than any other country in 2006, while China recently surpassed the European countries that are members of the Organization for Economic Cooperation and Development (OECD) as the world's second largest emitter. On a per capita basis, the United States and Canada rank first and second in carbon dioxide emissions. Between 1990 and 2006, carbon dioxide emissions grew substantially in both China and India.

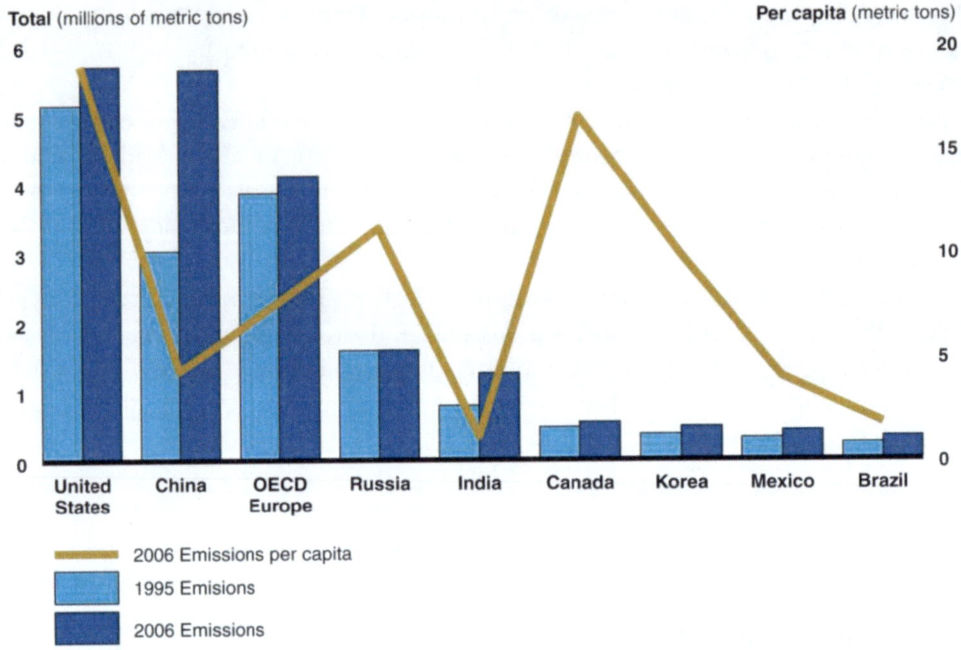

Source: International Energy Agency data on carbon dioxide emissions from fuel consumption

Figure 1. Data on Global Carbon Dioxide Emissions

The Greenhouse Effect

Greenhouse gases, such as carbon dioxide, methane, nitrous oxide, and other gases, increase temperatures by trapping heat that would otherwise escape the earth's atmosphere. The heat-trapping effect, known as the greenhouse effect, moderates atmospheric temperatures, keeping the earth warm enough to support life. Although each unit of non-carbondioxide greenhouse gas generally has a greater warming effect than each unit of carbon dioxide, carbon dioxide is the most important greenhouse gas because of its significant volume. The potency of other greenhouse gases may be expressed in terms of their warming potential relative to carbon dioxide, or their carbon dioxide equivalent.

Estimated Growth in Emissions

According to the IPCC, in 2004, developed countries, including the United States, constituted 20 percent of the global population, but were responsible for nearly half of global greenhouse gas emissions. However, the IPCC projects that between 2000 and 2030, two-thirds to three-quarters of the projected increase in global carbon dioxide emissions will occur in developing countries. The IPCC also projects that compared with 2000, global greenhouse gas emissions will increase between 25 percent and 90 percent by 2030 in the absence of policies to reduce greenhouse gas emissions worldwide.

International Negotiations and Emissions Pricing

Greenhouse gas emissions reductions in all major countries will be required to stabilize greenhouse gas concentrations at a level that would prevent dangerous climate change. In 1992, 192 countries, including the United States, joined the United Nations Framework Convention on Climate Change (UNFCCC), an international treaty to consider what may be done to reduce and adapt to global warming. In 1998, the United States signed the Kyoto Protocol, an international agreement to specifically limit greenhouse gas **emissions**. The United States is not bound by the protocol's terms because it was not ratified by the Senate. Both the UNFCCC and the Kyoto protocol operate under the principal of "common but differentiated responsibilities" to reflect the agreement that each nation must contribute to addressing climate change, but that the magnitude of its efforts should differ according to national circumstances. For example, under the protocol, only **Annex I countries** have quantified emission reduction obligations. To negotiate the next round of international commitments, parties to the UNFCCC agreed to the "Bali Action Plan" in December 2007.

Countries can take varying approaches to reducing greenhouse gas emissions. Since energy use is a significant source of greenhouse gas emissions, policies designed to increase energy efficiency or induce a switch to less **greenhouse gas-intensive** fuels, such as from coal to natural gas, can reduce emissions in the short term. In the long term, however, major technology changes will be needed to establish a less carbon-intensive energy infrastructure. To that end, a U.S. policy to mitigate climate change may require facilities to achieve specified reductions or employ a market-based mechanism, such as establishing a price on emissions.

The Kyoto Protocol

The Kyoto Protocol, adopted in Kyoto, Japan, and ratified by more than 180 countries, sets legally binding emissions targets for 37 industrialized countries and the European Community (Annex I countries).

The Bali Action Plan

The year 2012 will mark the end of the first commitment period of the Kyoto Protocol. In Bali, Indonesia, in 2007, countries agreed to the Bali Action Plan, which defined a process and timeframe to enable the full effective and sustained implementation of the Convention through long-term cooperative action up to and beyond 2012. According to the Bali Action Plan, the negotiations process is scheduled to conclude in December 2009 at a major summit in Copenhagen, Denmark.

Existing Emissions Pricing Systems

Greenhouse gas emissions pricing has been implemented in several other cases, including a cap-andtrade system in the European Union (EU) and carbon tax systems in Sweden, Finland, Norway, and the Netherlands. The world's largest emissions pricing system, the European Union Emissions Trading Scheme (EU ETS), covers CO_2 emissions from more than 11,000 energy-intensive installations. The EU ETS began with a pilot phase that ran from 2005 to 2007. Its second trading period is currently under way. For more information see GAO-09-151.

Proposed Legislation Includes Two Options for Domestic Emissions Pricing Systems

Through greenhouse gas **emissions pricing**, governments create an incentive for parties to lower their greenhouse gas emissions by placing a cost on the emissions. Proposed U.S. legislation includes two basic approaches to greenhouse gas emissions pricing: a **carbon tax system and a cap-and-trade system.** A carbon tax system would impose a fee on emissions of **greenhouse gases**. Under a cap-and-trade system, the government would cap greenhouse gas emissions, limiting the total quantity of emissions from regulated sources under the system. The government would also issue **allowances**, or permits to emit greenhouse gases, equal to the overall cap or quota. Regulated sources would be required to hold enough allowances to cover their emissions, and allowances could be bought and sold as needed to meet the system's requirements in the least expensive manner.

Several bills to implement emissions pricing in the United States have been introduced in the 110th and the 111th Congresses. These bills have included both cap-and-trade and carbon tax proposals. Some of the proposed legislation also include measures intended to limit potentially adverse impacts on the international **competitiveness** of domestic firms.

1. Estimating Industry Effects

Introduction

In this section, we present information on what is known about how U.S. **emissions pricing** could potentially affect industry **competitiveness**. Importantly, estimating the effects of domestic emissions pricing for industries in the United States is complex. For example, if the United States were to regulate greenhouse gas emissions, production costs could rise for many industries and could cause output, profits, or employment to fall. However, the magnitude of these potential effects is likely to depend on the **greenhouse gas intensity** of industry output and on the domestic emissions price, which is not yet known, among other factors. Additionally, if U.S. climate policy were more stringent than in other countries, some domestic industries could experience a loss in international competitiveness. Within these industries, adverse competitiveness effects could arise through an increase in imports, a decrease in exports, or both.

To estimate the potential economic effects of U.S. emissions pricing on domestic industries, models require key assumptions, which cause estimates to vary. In this section, we review two economic models that specifically estimate competitiveness effects for U.S. industries. In each, larger adverse effects occur for those U.S. industries that are both relatively energy and trade intensive. We also present information on the potential for adverse international competitiveness effects to be a source of **carbon leakage**.

U.S. Emissions Pricing Effects on Production Costs and Output

For regulated sources, greenhouse gas emissions pricing would increase the cost of releasing **greenhouse gases**. As a result, it would encourage some of these sources to reduce their emissions, compared with **business-as-usual**. Under domestic emissions pricing, production costs for regulated sources could rise as they either take action to reduce their emissions or pay for the greenhouse gases they release. Cost increases are likely to be larger for production that is relatively greenhouse gas intensive, where greenhouse gas intensity refers to emissions per unit of output. Cost increases may reduce industry profits, or they may be passed on to consumers in the form of higher prices. To the extent that cost increases are passed on to consumers, they could demand fewer goods, and industry output could fall.

While emissions pricing would likely cause production costs to rise for certain industries, the extent of this rise and the resulting impact on industry output are less certain due to a number of factors. For example, the U.S. emissions price and the emissions price in other countries are key variables that will help to determine the impact of emissions pricing on domestic industries. However, future emission prices are currently unknown. Furthermore, to the extent that emissions pricing encourages technological change that reduces greenhouse gas intensity, potential adverse effects of emissions pricing on profits or output could be mitigated for U.S. industries. Additionally, policy options such as output-based rebates could also offset some of the added costs of emissions pricing for some firms but involve tradeoffs, such as possibly increasing costs for firms that do not receive them.

Estimated Effects from Economic Models of Emissions Pricing

Several studies by U.S. agencies and experts have used models of the economy to simulate the effects of emissions pricing policy on output and related economic outcomes. These models generally find that emissions pricing will cause output, profits or employment to decline in sectors that are described as energy intensive, compared with business-as-usual. In general, these studies conclude that these declines are likely to be greater for these industries, as compared to other sectors in the economy. However, some research suggests that not every industry is likely to suffer adverse effects from emissions pricing. For example, a long-run model estimated by **Ho, Morgenstern, and Shih (2008)** predicts that some U.S. sectors, such as services, may experience growth in the long run as a result of domestic emissions pricing. This growth would likely be due to changes in consumption patterns in favor of goods and services that are relatively less greenhouse gas intensive.

Potential international competitiveness effects depend in part on the stringency of U.S. climate policy relative to other countries. For example, if domestic greenhouse gas emissions pricing were to make emissions more expensive in the United States than in other countries, production costs for domestic industries would likely increase relative to their international competitors, potentially disadvantaging industries in the United States. As a result, some domestic production could shift abroad, through changes in consumption or investment patterns, to countries where greenhouse gas emissions are less stringently regulated. For example, consumers may substitute some goods made in other countries for some goods made domestically. Similarly, investment patterns could shift more strongly in favor of new capacity in countries where greenhouse gas emissions are regulated less stringently than in the United States.

Economists such as **Aldy and Pizer (2009)** have defined adverse international competitiveness effects as the part of a decline in domestic output that is due to a shift of production abroad in response to emissions pricing, or the part of a decline in domestic output that not also matched by a decline in domestic consumption. For example, if domestic output were to fall by more than domestic consumption, the difference could be explained by changes in trade patterns through a reduction in exports, an increase in imports, or both.

Factors That May Affect International Competitiveness

Stakeholders and experts have identified two criteria, among others, that are important in determining potential vulnerability to adverse competitiveness effects: **trade intensity** and **energy intensity**. Trade intensity is important because international competitiveness effects arise from changes in trade patterns. For example, if climate policy in the United States were more stringent than in other countries, international competition could limit the ability of domestic firms to pass increases in costs through to consumers. Energy intensity is important because the combustion of fossil fuels for energy is a significant source of greenhouse gas emissions, which may increase production costs under emissions pricing. International competitiveness effects may also depend on other factors, including transportation costs and access to markets for natural resources, capital, or labor.

Industries that are both relatively trade intensive and energy intensive include primary metals, nonmetallic minerals, paper products, and chemicals. These four industries provided around 4.5 percent of total U.S. output in 2007.

Legislation passed in June 2009 by the House of Representatives, H.R. 2454, 111th Cong. (2009), uses the criteria of trade intensity and energy intensity or greenhouse gas intensity, among others, to determine eligibility for the Emission Allowance Rebate Program that is part of the legislation. H.R. 2454 specifies how to calculate the two criteria. Trade intensity is defined as the ratio of the sum of the value of imports and exports within an industry to the sum of the value of shipments (output) and imports within the industry. Energy intensity is defined as the industry's cost of purchased electricity and fuel costs, or energy expenditures, divided by the value of shipments (output) of the industry.

The energy intensity measure specified in this legislation, however, might not be an ideal measure of vulnerability. First, the relationship between energy expenditures and emissions from energy use is indeterminate. For example, energy expenditures depend on the price of fuel and on the quantity of fuel used, while emissions from energy use depend on the quantity and type of fuel used. Additionally, an increase in energy prices may cause expenditures on energy to rise and energy use to fall, which could also cause emissions from energy use to fall. Second, the energy intensity measure might not reflect the extent to which substitutes for energy are available. For example, regulated sources may be able to reduce their energy intensity or switch to energy sources that are less greenhouse gas intensive. To the extent that less greenhouse gas intensive substitutes are available, some industries could reduce the effect of emissions pricing on their costs and on their competitiveness.

Key Assumptions of Models That Estimate Competitiveness Effects

Key assumptions made by economic models of emissions pricing include, among others, the domestic emissions price, the emissions price in other countries, the level of industry aggregation, and the time horizon during which producers may adjust their production methods or energy sources in response to emissions pricing.

- *Domestic emissions price:* The domestic emissions price could affect the size of a potential increase in production costs from emissions pricing. In particular, increases in production costs are expected to be larger when the domestic emissions price is greater. Under a **carbon tax system**, the domestic emissions price would equal the tax per unit of emissions. Under a cap-and-trade system, the domestic emissions price would equal the market-clearing allowance price, at which the quantity of **allowances** supplied equals the quantity of allowances demanded.
  The domestic emissions price could also depend on cost containment features of an emissions pricing system, such as banking, borrowing or **carbon offsets**, among other features.

- *Emissions price in other countries*: Together with the domestic emissions price, the emissions price in other countries will help to determine the relative stringency of climate policy in the United States. When the difference between the domestic emissions price and a lower emissions price in other countries is larger, U.S. industries may be more likely to experience adverse competitiveness effects from domestic **emissions pricing**. Under a cap-and-trade system, differences in the allowance price between two countries may dissipate if international trading of allowances is allowed.

- *Level of aggregation*: The range of estimated effects may depend on the level of industrial or sectoral aggregation used. For example, estimates for highly aggregated sectors could mask more extreme variation at the more narrowly aggregated sub-industry level.

- *Time horizon*: In models of economic activity, the amount of time considered after policy implementation may affect model results. For example, the time horizon may affect the ability of firms to pass through added costs to consumers in the form of higher prices, change their production processes, or develop and adopt new technologies. A longer time horizon could also allow for investments in new capacity.

- *Policy Scenarios*: Model results may depend in part upon the range of policy scenarios incorporated. For example the incorporation of trade measures and output-based rebates may affect the results.

Table 1. Estimates of Emissions Pricing Effects on Energy-Intensive Manufacturing Industries in the United States

	Study	
	Aldy and Pizer (2009)	**Ho, Morgenstern, and Shih (2008)**
Estimated decline in industry output due to emissions pricing (range of estimates)	1.0% – 4.3%	0.91% - 1.30%
Part of estimated decline in industry output due to adverse international competitiveness effects (range of estimates)	0.3% – 1.8%	0.42% - 0.68%
Domestic emissions price (per ton of carbon dioxide)	$15	$10
Emissions price in other countries (per ton of carbon dioxide)	$0	$0
Time horizon	1 year to a few years	Long run
Number of industries in model	400+	21
Number of industries for which results are reported by authors	55 energy-intensive	21
Number of industries included in the ranges of estimates reported above	55	3 most energy-intensive
Incorporation of trade measure and output-based rebate policies	no	no

Source: GAO analysis of results reported by Aldy and Pizer (2009) and Ho, Morgenstern, and Shih (2008). Note: The number of industries in both reflects, in part, the level of industry aggregation. While Aldy and Pizer (2009) examine more than 400 industries at a more narrowly aggregated level, they report estimates only for 55 energy-intensive industries among the more than 400 industries they examine. Ho et al. (2008) examine industries at a more aggregated level and report estimates for all 21 manufacturing and nonmanufacturing sectors that they consider. Ho et al. (2008) also report estimates of the effects of emissions pricing on industry output for multiple time horizons, but they report both production and consumption effects—both of which are necessary to compute competitiveness effects—only for their long run estimates. For the study by Aldy and Pizer (2009), the estimated decline in industry output and the part that is due to competitiveness effects are reported as a percentage of industry output. For the study by Ho et al. (2008), the estimated decline in industry output and the part that is due to competitiveness effects are reported as a share of sectoral consumption.

Estimates of Competitiveness Effects are Uncertain

Table 1 presents results from two studies that attempt to quantify the potential competitiveness effects on domestic industries from emissions pricing. For each study, a range of estimates is reported for two effects: (1) the potential decline in domestic, energy-intensive manufacturing output that is due to U.S. emissions pricing and (2) the part of this decline that is due to adverse international competitiveness effects. The adverse competitiveness effects are computed as the difference between the production and consumption impacts of emissions pricing. For example, if production declines by 3.7 percent and consumption declines by 2.9 percent as a share of production, the adverse competitiveness effect equals the -0.8 percent change in production that is not matched by a change in domestic consumption.

In both studies, estimates of competitiveness effects are generally greater for industries that are relatively trade and energy intensive.

The estimated impacts depend on key assumptions of the models used to generate these results. Both studies assume unilateral action by the United States, or an emissions price of $0 per ton of carbon dioxide in the rest of the world, which may overstate the estimates of adverse competitiveness effects for U.S. industries to the extent that other countries also regulate greenhouse gas emissions.

Aldy and Pizer's (2009) model allows for some adjustments and is consistent with a time horizon of 1 year to a few years. Ho et al. (2008) generate their long-run estimates using a model that allows for more adjustments.

International Competitiveness Effects and Carbon Leakage

Reducing carbon emissions in the United States could result in carbon leakage through two potential mechanisms. First, if domestic production were to shift abroad to countries where greenhouse gas emissions are not regulated, emissions in these countries could grow faster than expected otherwise. Through this mechanism, some of the expected benefits of reducing emissions domestically could be offset by faster growth in emissions elsewhere, according to **Aldy and Pizer (2009)**.

Carbon leakage may also arise from changes in world prices that are brought about by domestic emissions pricing. For example, U.S. emissions pricing could cause domestic demand for oil to fall. Because the United States is a relatively large consumer of oil worldwide, the world price of oil could fall when the U.S. demand for oil drops. The quantity of oil consumed by other countries would rise in response, increasing greenhouse gas emissions from the rest of the world. These price effects may be a more important source of carbon leakage than the trade effects described above. See **Fischer and Fox (2009)** and **EPA (2009)**.

2. Potentially Vulnerable Industries

Introduction

In this section, we provide some information on certain U.S. industries that, by paying a price on greenhouse gas emissions, could potentially lose **competitiveness** compared with foreign industries without comparable climate policies. Among many factors, two key indicators of potential vulnerability to adverse competitiveness effects are an industry's **energy intensity** and **trade intensity**. Proposed U.S. legislation specifies that: (a) either an energy intensity or greenhouse gas intensity of 5 percent or greater; and (b) a trade intensity of 15 percent or greater be used as criteria to identify industries for which trade measures or rebates would apply. Since data on greenhouse gas intensity is less complete, we focus our analysis on industry energy intensity. Most of the industries that meet these criteria are in primary metals, nonmetallic minerals, paper, and chemicals, yet there is significant variation in specified vulnerability characteristics among different product groups ("sub-industries"). The following pages include examples of this variation, as well as information on the type of energy used and location of import and export markets. Data shown are for the latest year available, with additional discussion of proposed vulnerability criteria in appendix I and industry information in appendix II.

Energy and trade intensities of the four manufacturing industries, in the aggregate, generally meet both vulnerability criteria in proposed legislation. As shown along the right axis in figure 2, nonmetallic minerals is the most energy intensive at 6.1 percent in 2006. As shown along the left axis, chemicals is the most trade intensive, at 45 percent in 2007. While chemicals does not have an energy intensity of 5 percent or greater -- and is located outside of the shaded area— several subindustries within chemicals meet that vulnerability criterion.

Together, these four industries provided 23 percent of total U.S. manufacturing output in 2007 and had trade flows of about $500 billion. As shown by the size of the column, chemicals is the largest industry, with output of $664 billion in 2007. Accordingly, chemicals also accounted for the largest share of carbon dioxide emissions from manufacturing, at 22 percent in 2002. However, the estimated greenhouse gas intensity of chemical products— the level of carbon dioxide emitted per unit of energy—was less than that for both primary metals and nonmetallic minerals in 2002.

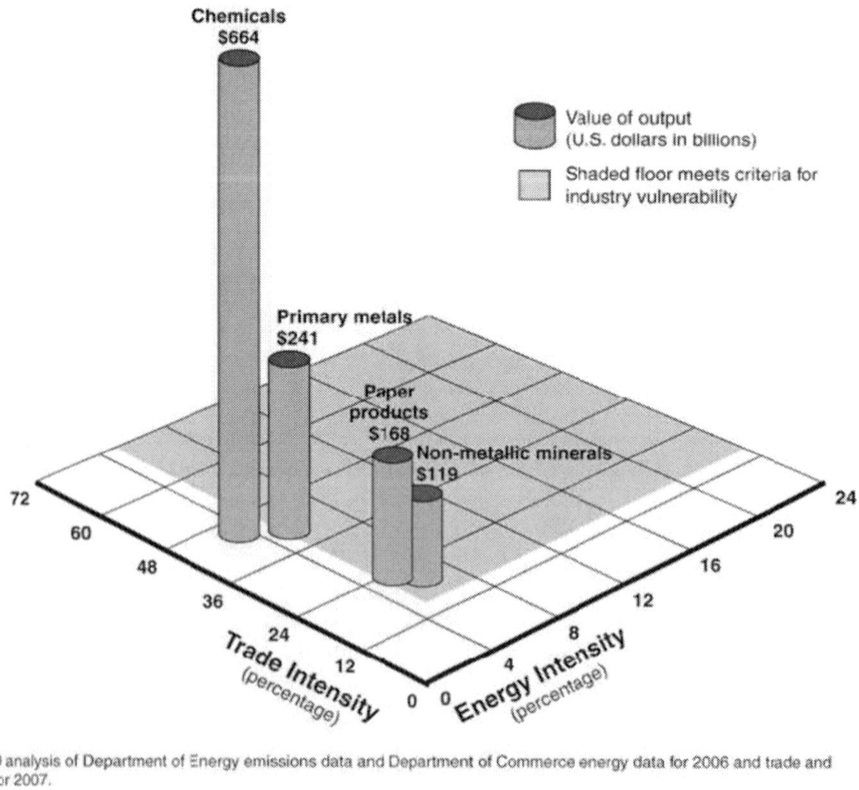

Source: GAO analysis of Department of Energy emissions data and Department of Commerce energy data for 2006 and trade and output data for 2007.

Figure 2. Energy and Trade Intensity Are Indicators of Vulnerability.

2.1. Primary Metals Examples

As shown by sub-industry examples in figure 3, energy and trade intensities differ within primary metals. For example, primary aluminum meets the vulnerability criteria with an energy intensity of 24 percent and trade intensity of 62 percent. Ferrous metal foundries meets the energy intensity criteria, but not the trade intensity criteria. Steel manufacturing—products made from purchased steel—and aluminum products fall short of both vulnerability criteria. Iron and steel mills has an energy intensity of 7 percent and a trade intensity of 35 percent and is by far the largest sub- industry example, with a 2007 value of output of over $93 billion.

Among the primary metals sub-industry examples, types of energy used also vary. Iron and steel mills use the greatest share of coal and coke, and steel manufacturing and ferrous metal foundries use the greatest proportion of natural gas. Since coal is more carbon intensive than natural gas, sub-industries that rely more heavily on coal could also be more vulnerable to competitiveness effects. The carbon intensity of electricity, used heavily in the production of aluminum, will also vary on the basis of source of energy used to generate it and the market conditions where it is sold. Data shown for "aluminum" include primary aluminum and aluminum products and net electricity is the sum of net transfers plus purchases and generation minus quantities sold.

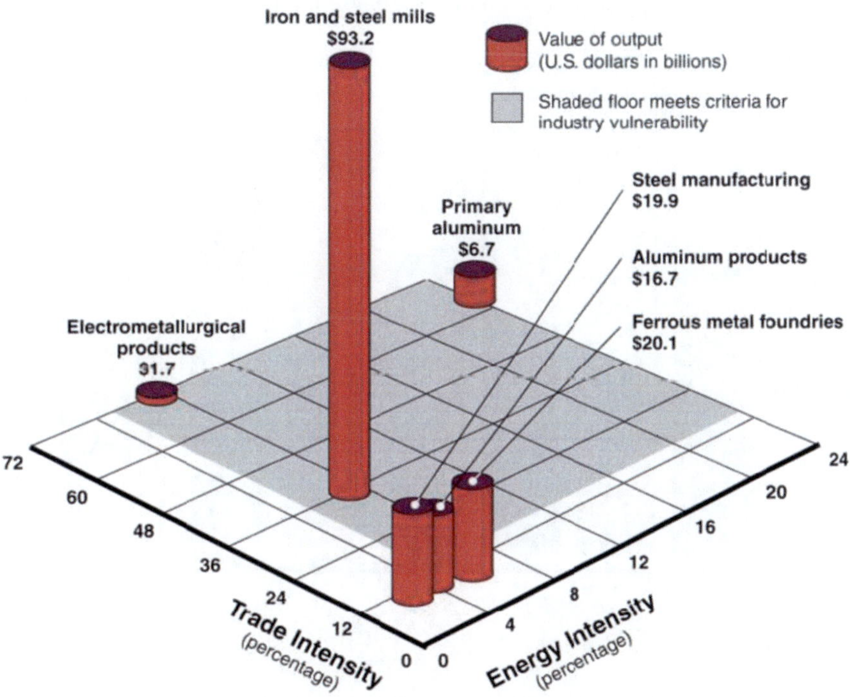

Source: GAO analysis of Department of Commerce energy data for 2006 and trade data for 2007.

Figure 3. Energy and Trade Intensity Indicators Vary by Sub- Industry

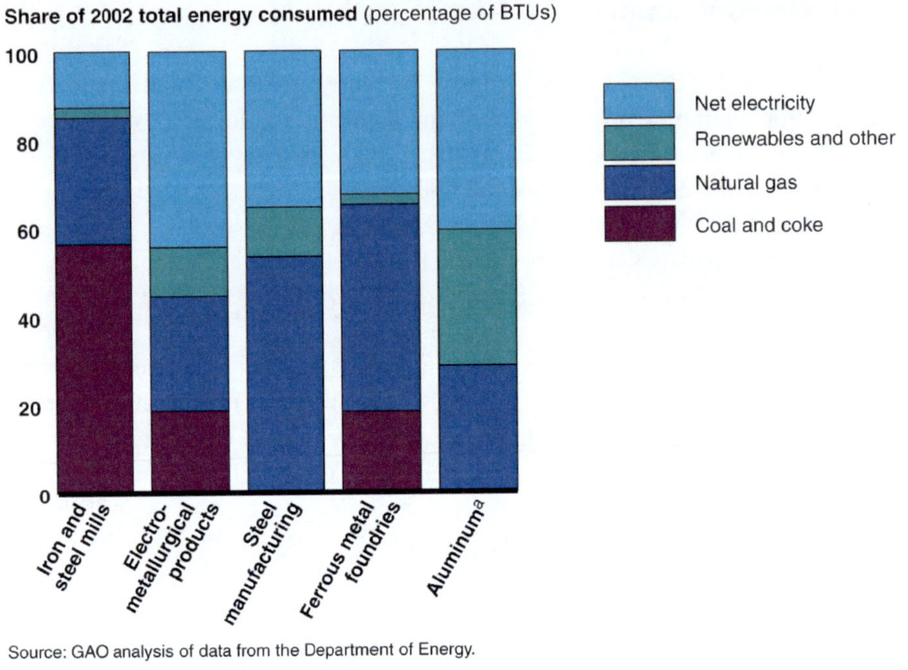

Figure 4. Type of Energy Used Is Important for Carbon Intensity

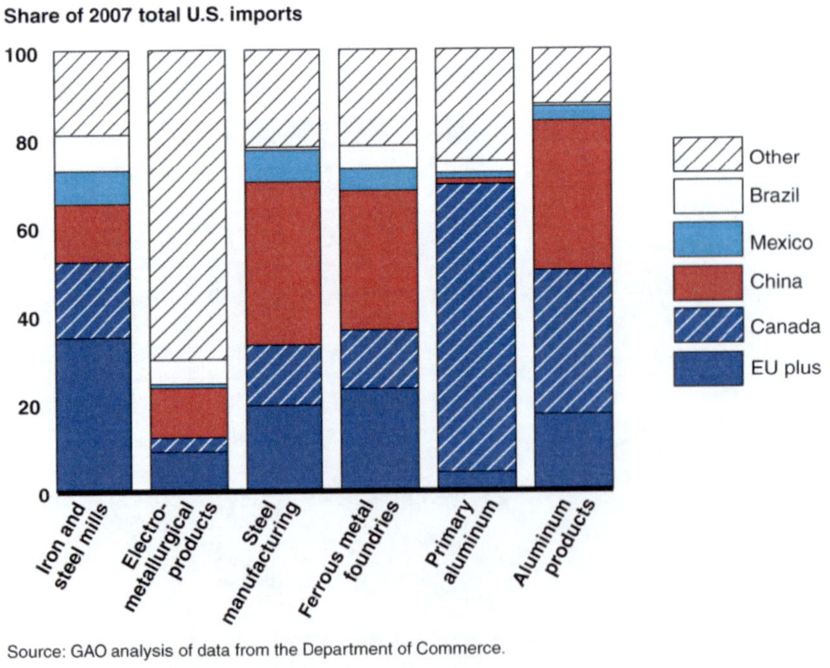

Figure 5. Source of Imports Is Important for Trade Intensity.

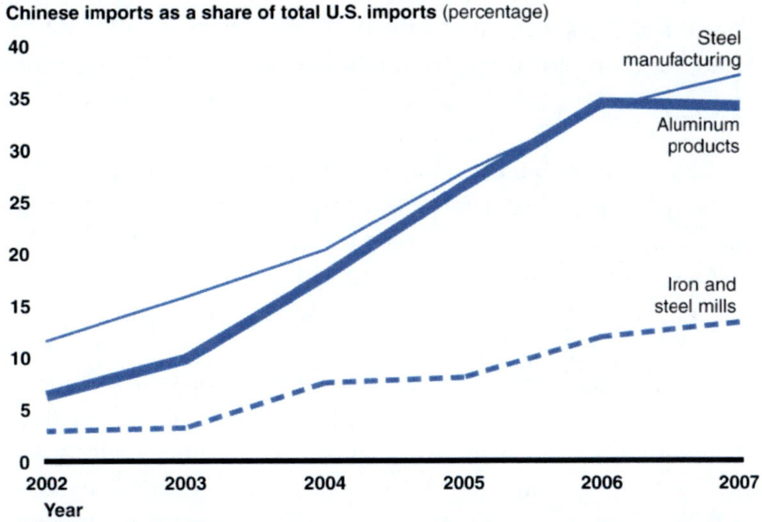

Total U.S. imports by value (U.S. dollars in millions)						
Year	**2002**	**2003**	**2004**	**2005**	**2006**	**2007**
Iron and steel mills	$12,558	10,808	23,355	25,131	33,060	30,445
Steel manufacturing	$1,110	1,184	1,792	1,884	1,916	1,822
Aluminum products	$498	549	718	913	1,209	1,183

Source: GAO analysis of data from the Department of Commerce.

Figure 6. Sub-Industries With Growing Share of Imports from China.

Industry vulnerability may further vary depending on the share of trade with countries that do not have carbon pricing. To illustrate this variability, figure 5 provides data on the share of imports by source, since imports exceed exports in each of the primary metals examples. As shown, while primary aluminum is among the most trade intensive, the majority of imports are from Canada, an Annex I country with agreed emission reduction targets. For iron and steel mills, over one- third of imports are from the EU and other Annex I countries, not including Canada ("EU plus"). However, for iron and steel mills, almost 30 percent of imports are also from the non-Annex I countries of China, Mexico, and Brazil. While less trade intensive, steel manufacturing and aluminum products each has greater than one- third of imports from China alone.

Adverse competitiveness effects from emissions pricing could increase the already growing share of Chinese imports that exists in some of the sub-industries. Among the examples, iron and steel mills, steel manufacturing, and aluminum products exhibit a growing trade reliance on Chinese imports since 2002. This trend has largely been driven by lower labor and capital costs in China and, according to representatives from the steel industry, China has recently been producing 50 percent of the world's steel.

2.2. Nonmetallic Minerals Examples

Among nonmetallic minerals sub- industry examples, cement has the highest energy intensity, at 16 percent, and glass and clay building materials have the highest trade intensity, at 36 percent and 35 percent, respectively. While, concrete is the largest sub-industry, with a value of output of over $30 billion in 2007, it fails to meet both of the vulnerability criteria and thus lies outside of the shaded floor area. Lime and gypsum meets the energy intensity criteria but fails to meet the trade intensity criteria.

Cement and lime rely on coal and coke sources for about 60 percent of energy consumed, while glass and mineral wool rely on natural gas for over 65 percent of energy consumed. Since coal is more carbon intensive than natural gas, sub-industries that rely more heavily on coal could also be more vulnerable to competitiveness effects. Data on energy use by the sub-industry examples excluded from the chart are not currently reported by the Department of Energy.

For the nonmetallic mineral examples, total imports exceed total exports. As shown in figure 9, glass has the largest share of imports from China, at 26 percent. Another 17 percent of glass imports are sourced from Mexico. For the other trade- intensive sub-industries of cement, clay building material, and mineral wool, about 30 percent of imports are from China, Mexico, and Brazil. Conversely, nearly 60 percent of lime and gypsum imports are from Canada.

The one sub-industry where exports comprise more than 60 percent of trade flows is concrete, with 30 percent of concrete exports sold to markets in Mexico and the Middle East.

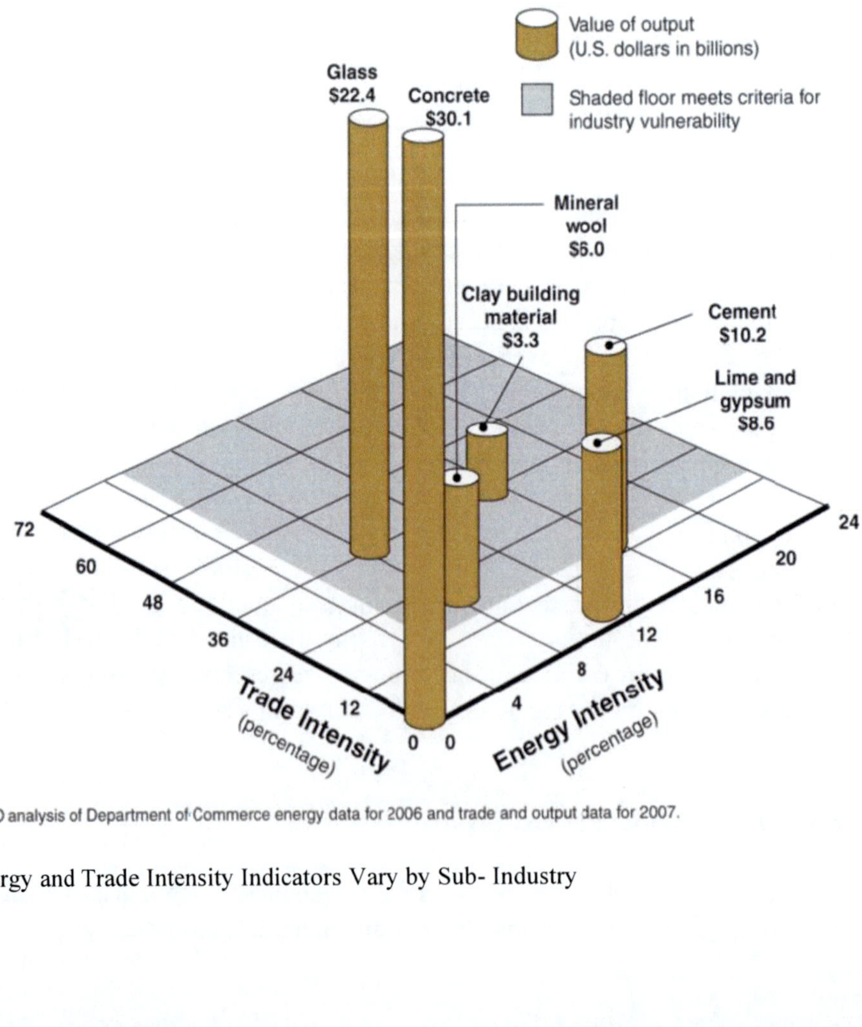

Figure 7. Energy and Trade Intensity Indicators Vary by Sub- Industry

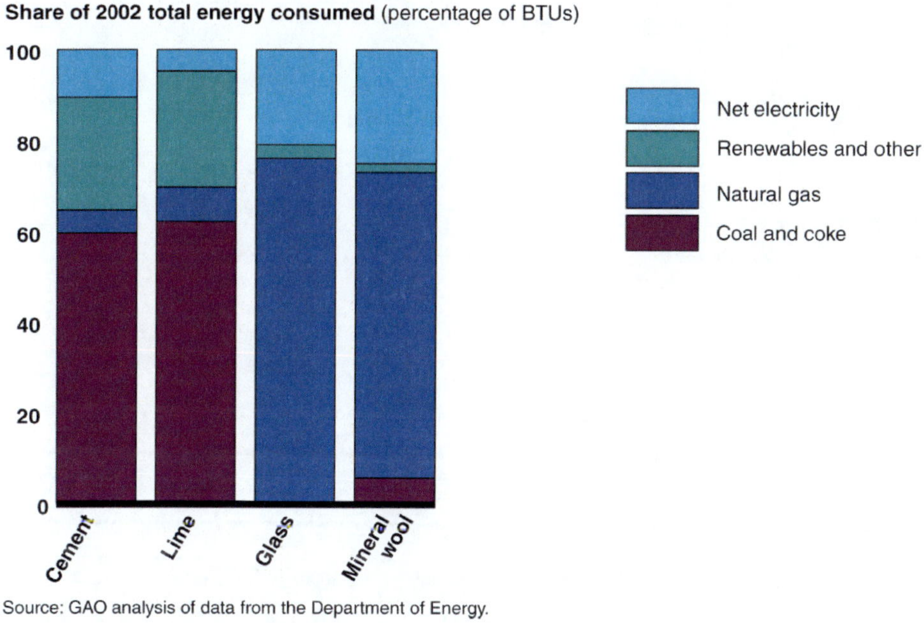

Figure 8. Type of Energy Used Is Important for Carbon Intensity

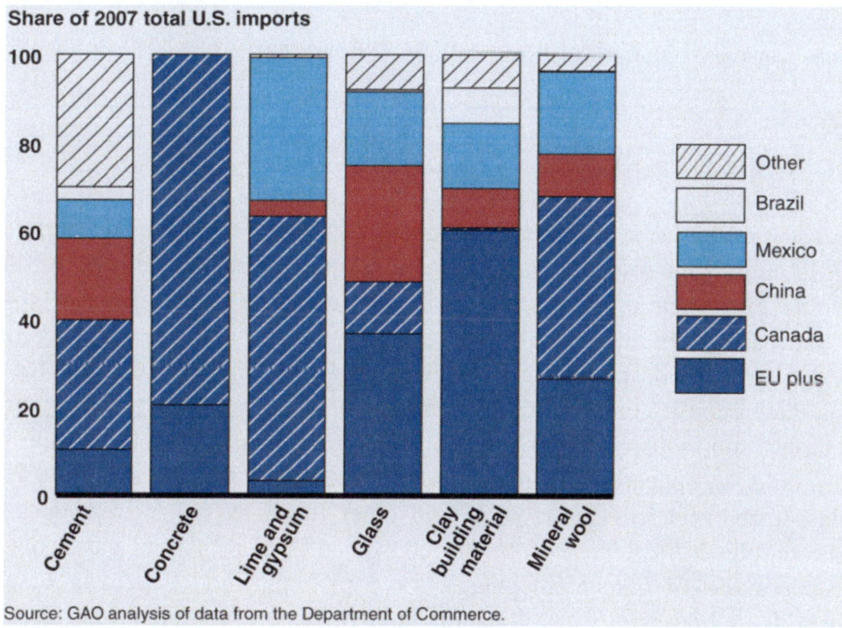

Figure 9. Source of Imports Is Important for Trade Intensity.

Since 2002, a trend of increasing Chinese imports as a share of total imports is evident for glass and mineral wool. While the share of cement imports from China declined from 2006 to 2007, reliance on Chinese imports has also grown relative to 2002. According to representatives of the U.S. cement industry, Chinese production costs are between 20 percent and 40 percent of those in the United States.

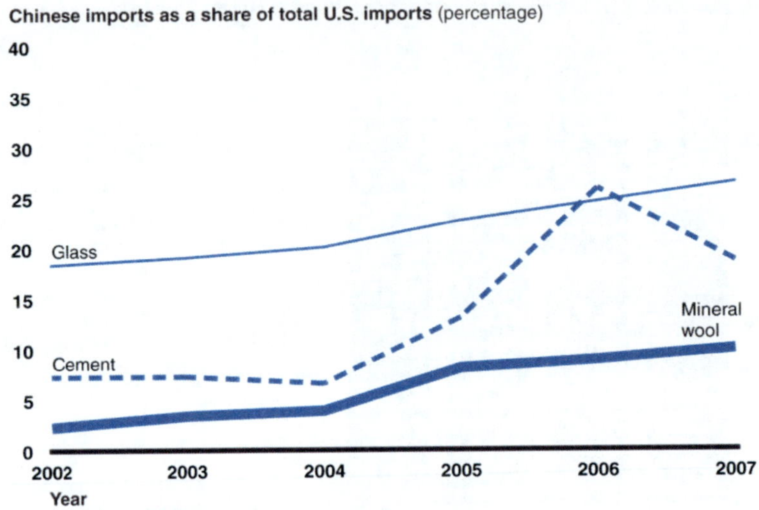

Total U.S. imports by value (U.S. dollars in millions)						
Year	2002	2003	2004	2005	2006	2007
Cement	$938	924	1,140	1,567	1,837	1,325
Glass	$4,293	4,436	4,947	5,125	5,491	5,726
Mineral wool	$332	360	448	510	577	489

Source: GAO analysis of data from the Department of Commerce.

Figure 10. Sub-Industries With Growing Share of Imports from China.

2.3. Paper Products Examples

Sub-industry examples within paper products also show variation in energy and trade intensity. Pulp mills is the most trade intensive, at 98 percent. Paper mills is the largest sub-industry, with a 2007 value of output of over $51 billion and its trade intensity is 30 percent. Pulp mills and paper mills each has an energy intensity of 8 percent and, thus, meet both vulnerability criteria. While paperboard mills has an energy intensity of 12 percent, it does not meet the trade intensity criteria. Paperboard containers fails to meet both criteria but was the second largest sub-industry in 2007, with a value of output of over $47 billion.

In paper mills and paperboard mills, the largest share of energy consumed is from the relatively less carbon-intensive renewable and other (such as fuel oil and steam) energy sources. An additional 20 percent of energy consumed is from natural gas, while coal and coke sources provide 14 percent of energy for paper mills and 9 percent of energy for paperboard mills. While energy use data for pulp mills is reported by the Department of Energy, consumption of coal and coke, specifically, is not publicly available, so percentage shares could not be computed. Data on energy use by paperboard containers are not currently reported.

For the paper products sub-industry examples, total imports exceed total exports. For each, the largest share of imports is from Canada, an Annex I country with agreed emission reduction targets. However, for the most trade-intensive example of pulp mills, 19 percent of imports are from Brazil. While paperboard containers has a trade intensity of only 6 percent, 27 percent of imports are from China.

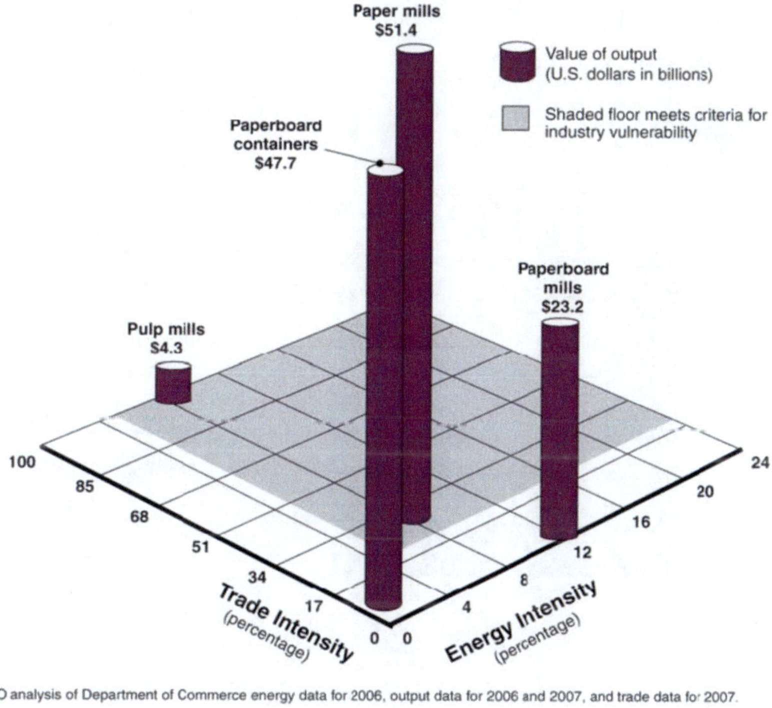

Figure 11. Energy and Trade Intensity Indicators Vary by Sub- Industry.

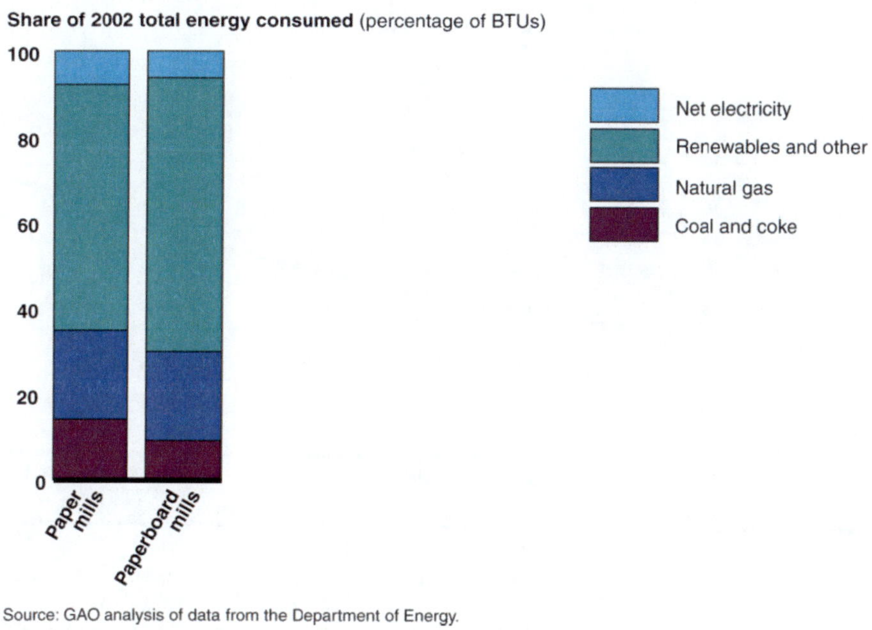

Figure 12. Type of Energy Used Is Important for Carbon Intensity.

Among the paper products subindustries, paperboard containers is the only example with a significant share of imports from China. While paperboard containers fails to meet the trade-intensity vulnerability criteria, emissions pricing could magnify the growing share of imports from China for that sub- industry. From 2002 to 2007, this share has grown from 11 percent to 27 percent of U.S. imports.

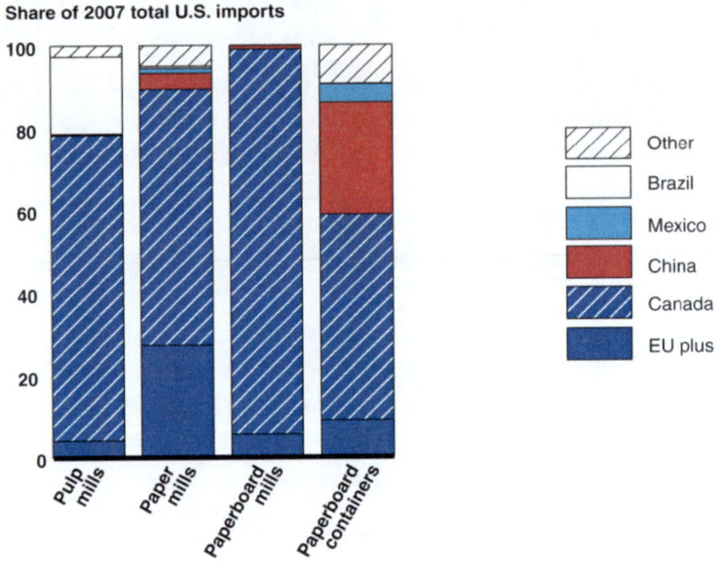

Source: GAO analysis of data from the Department of Commerce.

Figure 13. Source of Imports Is Important for Trade Intensity.

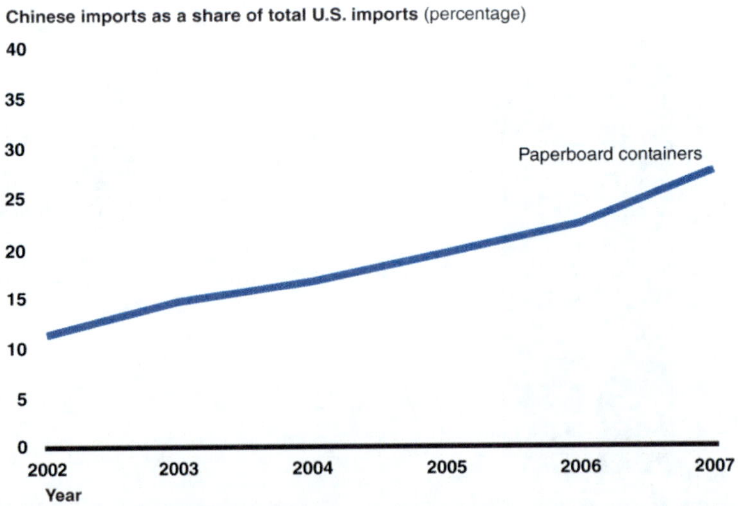

Total U.S. imports by value (U.S. dollars in millions)						
Year	**2002**	**2003**	**2004**	**2005**	**2006**	**2007**
Paperboard container	$790	875	926	1,033	1,187	1,275

Source: GAO analysis of data from the Department of Commerce.

Figure 14. Sub-Industries With Growing Share of Imports from China.

2.4. Chemicals Examples

Among chemicals examples, alkalies and chlorine is the most energy intensive, at 26 percent, and nitrogenous fertilizers is the most trade intensive, at 82 percent.Except for industrial gases, each of the examples meets the energy and trade intensity criteria for vulnerability. Industrial gases, the largest sub-industry example, with a 2007 value of output of almost \$9 billion, has an energy intensity of 15 percent but a trade intensity of only 5 percent. Not shown in figure 15 is petrochemicals, with a 2007 value of output of almost \$62 billion. The Department of Commerce does not publicly provide energy-intensity data for petrochemicals, though industry experts estimate it to be at above 5 percent but below 20 percent. The year 2007 trade data indicate a petrochemicals trade intensity of below 15 percent.

Carbon black relies mostly on renewable and other sources of energy that are relatively less carbon intensive than coal. Nearly all of the energy consumed by nitrogenous fertilizers is from natural gas. In chemical sub-industries, a portion of the carbon contained in the energy source is also sequestered in the product rather than emitted to the atmosphere. Data on certain energy uses by type for sub-industries not shown in figure 16 is not publicly available and percentage shares could not be computed.

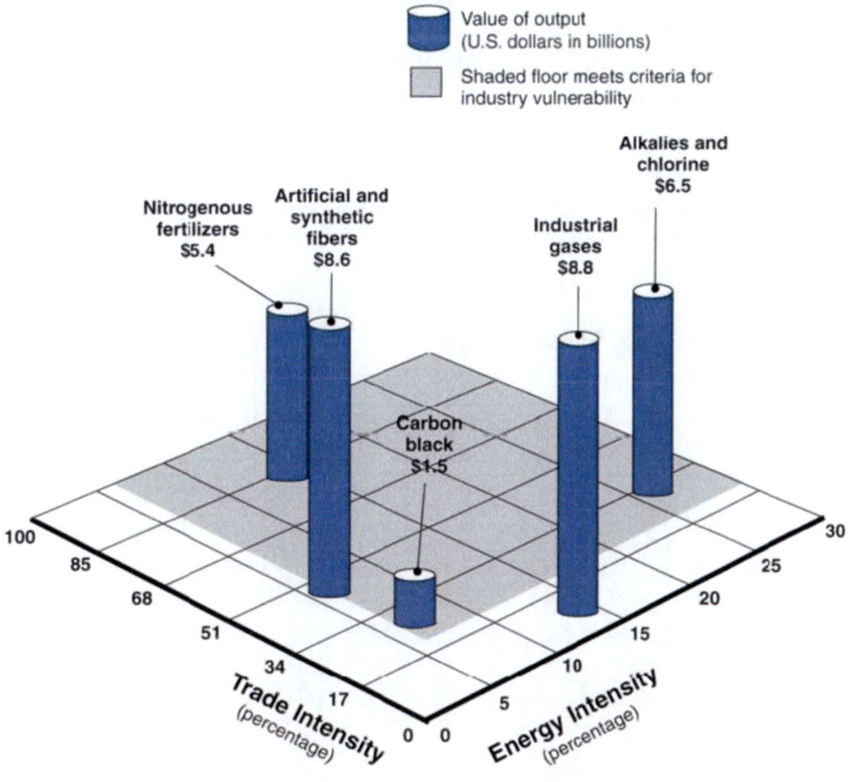

Source: GAO analysis of Department of Commerce energy data for 2006, output data for 2006 and 2007, and trade data for 2007.

Figure 15. Energy and Trade Intensity Indicators Vary by Sub- Industry

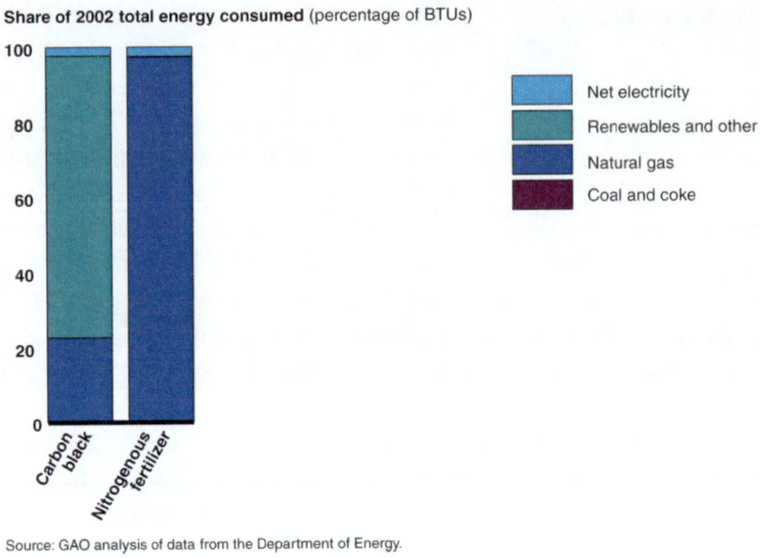

Figure 16. Type of Energy Used Is Important for Carbon Intensity.

For the chemicals examples, total imports exceed total exports. Nitrogenous fertilizers is the most trade intensive, and less than 40 percent of imports are from Annex I countries— Canada and EU Plus— while 40 percent of imports are from either Trinidad and Tobago or countries in the Middle East. Conversely, imports from Canada provide 60 percent of imports for carbon black. Imports from China, Mexico, and Korea account for 20 percent or more of imports in industrial gases and artificial and synthetic fibers.

Alkalies and chlorine is the only sub- industry example where exports comprise more than 60 percent of trade flows, with 15 percent to Brazil, 13 percent to Canada, and 12 percent to Mexico.

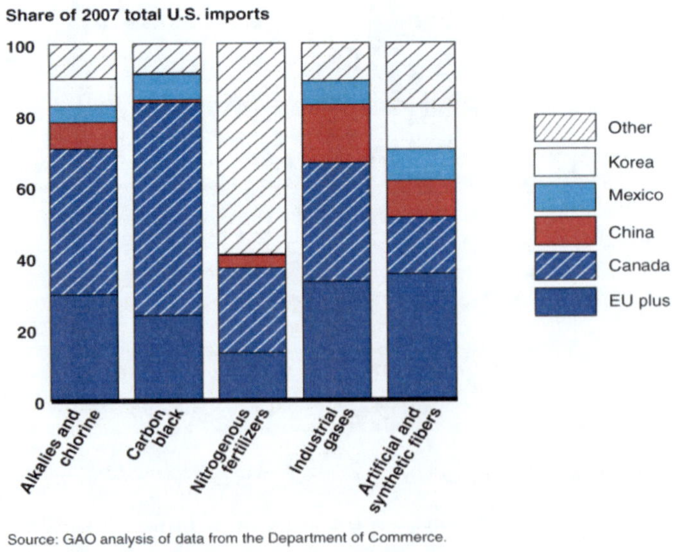

Figure 17. Source of Imports Is Important for Trade Intensity.

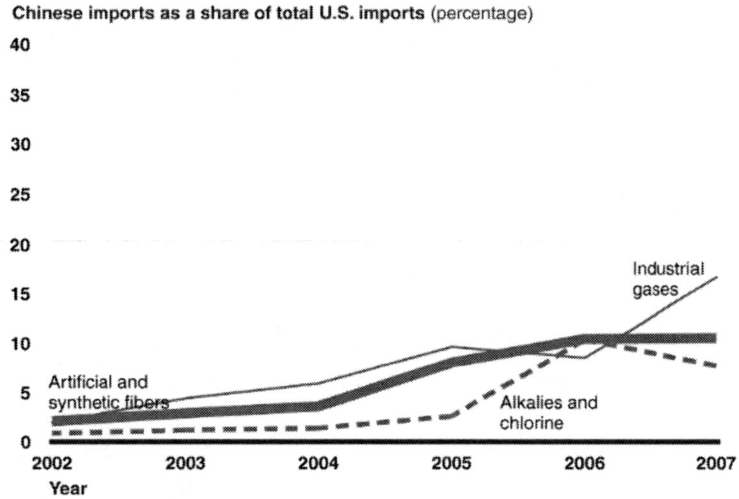

Total U.S. imports by value (U.S. dollars in millions)						
Year	2002	2003	2004	2005	2006	2007
Alkalies and chlorine	$160	207	253	454	460	400
Industrial gases	$125	128	148	160	160	176
Artifical and synthetic fibers	$1,626	1,644	1,814	2,230	2,322	2,471

Source: GAO analysis of data from the Department of Commerce.

Figure 18. Sub-Industries With Growing Share of Imports from China

From 2002 to 2007, the share of imports from China has grown for alkalies and chlorine, industrial gases, and artificial and synthetic fibers from a share of less than 2 percent to a share of between 8 percent and 16 percent. According to representatives from the chemical industry, growing imports from China are also evident for industry downstream products, such as plastics.

2.5. Summary of Industry Examples

This section provides data on energy and trade characteristics for subindustries within primary metals, nonmetallic minerals, paper, and chemicals. As shown by the data, characteristics that contribute to an industry's or subindustry's potential vulnerability to adverse competitiveness effects vary significantly among the examples provided. Additional variation would likely exist for factors not discussed, such as transportation costs and access to markets for natural resources, capital, and labor. As a result, the data provided are not sufficient to determine if one sub-industry is more vulnerable than another. Instead, the data suggest that an assessment of vulnerability that is based only on a sub-industry's energy and trade intensity may mask important differences in vulnerability among industries assessed.

- **Primary metals examples**: Primary aluminum is the most energy and trade intensive, yet the largest share of imports is from Canada. Iron and steel is the largest sub-industry that is both energy and trade intensive and also has the greatest reliance on coal and coke energy sources. Steel manufacturing and aluminum products fall short of meeting the trade- intensity criteria, but compared with the other examples, have the largest share of imports from China and the most pronounced trend of an increased import share from China since 2002.

- **Nonmetallic mineral examples**: Cement is the most energy intensive and has a relatively greater reliance on coal and coke energies. Glass is the most trade intensive and has the largest share of imports from China. Both cement and glass show an increasing share of Chinese imports since 2002. Concrete, while not meeting the trade-intensity criteria, exports more than it imports, with almost one-third of exports destined for markets in Mexico and the Middle East.

- **Paper examples**: Paper mills is the largest sub-industry example, although most imports are from Annex I countries. Almost one-fifth of pulp mill imports is from Brazil, although pulp mills is the smallest example that meets both energy and trade intensity criteria. Paperboard containers do not meet either vulnerability criteria, but show an increasing share of imports from China since 2002.

- **Chemicals examples**: Nitrogenous fertilizers is the most trade intensive, and 40 percent of imports are from Trinidad and Tobago or countries in the Middle East. Alkalies and chlorine are the most energy intensive and exports account for 79 percent of trade flows, of which, over one-fourth are to markets in Brazil and Mexico. Industrial gases do not meet either vulnerability criterion, but compared with the other examples, have the largest share of imports from China and the most pronounced trend of an increased import share from China since 2002.

3. Trade Measures

Introduction

In this section, we discuss the use of **trade measures** and **output-based rebates** to address potential **competitiveness** and environmental effects of a domestic **emissions pricing** system. Through trade measures, such as a **border tax adjustment** or a **border allowance requirement**, international competitors without comparable climate policies would pay for greenhouse gas emissions associated with their exports to the United States. Output-based rebates have been proposed as another type of measure to offset the costs of climate policy for sectors that are exposed to unregulated competition through payments based on a per-unit government rebate or tax credit, or a per-unit allocation of emissions **allowances**.

How a trade measure is ultimately designed will have important implications for the effectiveness of the overall measure in addressing industry and environmental effects. In terms of output based rebates, the extent and design of the rebates could help address the industry effects, but could also affect the costs to other industries.

In this section we provide explanations of key features of trade and output- based measures included in climate change legislation introduced between April 2007 and June

2009. We also provide information on a range of views regarding the value of trade and output-based measures as policy tools to address competitiveness and environmental concerns. Additionally, we provide information on the potential implementation challenges associated with implementing these measures.

American Electric and Power Border Allowance Requirement

Some of the climate change bills introduced between April 2007 and June 2009 included provisions for a border allowance requirement. A border allowance requirement proposal was developed by the American Electric Power (AEP) and trade union representatives with the International Brotherhood of Electric Workers (IBEW). The AEP/IBEW proposal was introduced in July 2007 in S. 1766, 110TH Cong. (2007) and was also included in various other legislative proposals.

3.1. Different Types

Trade measures have been proposed to address competitiveness and environmental effects associated with implementing a domestic emissions pricing system, such as a **cap-and-trade system** or a **carbon tax system**. Generally, in the context of a domestic emissions pricing system, trade measures would impose a cost or other requirement on energy-intensive imports from countries with weaker climate policies. Various types of trade measures exist for addressing competitiveness effects associated with a domestic emissions pricing system. Among the cap-and-trade and carbon tax legislative proposals we examined, **border tax adjustments** or border allowance requirements are the most frequently proposed type of trade measures.

Border Tax Adjustment

According to experts, a border tax adjustment is a levy applied by the federal government, on certain imported goods at the border. Key features of a border tax adjustment include the following:

- *Tax proportionate to the imports' embedded carbon*: Generally, the tax applied to the import would be based on the carbon emissions associated with the production of the imported good.
- *Tax equivalent to domestic compliance costs*: Generally, the federal government would charge imported goods the equivalent of what a domestic producer of a similar good would pay to comply with a domestic emissions pricing system.

Border Allowance Requirement

According to experts, a border allowance requirement is a measure that would require importers to purchase **allowances** from the federal government prior to importing goods into the United States. Generally, key features of a border allowance requirement include the following:

- *Establishes a separate pool of allowances*: In some of the cap-and-trade bills proposed in 2008, the federal government would establish a reserve of allowances separate from allowances established in the domestic capand-trade system. The border allowances that importers would be required to submit would come from this reserve allowance pool.

- *Comparable action determination*: Importers of goods from foreign countries that do not take "comparable action" to the United States to limit greenhouse gas emissions, would be required to submit allowances to accompany exports to the United States of the covered good.

- *Documentation at the border*: Prior to entering the United States, importers would have to purchase allowances from the federal government, and submit a written declaration at the border stating that the good is accompanied by the required number of allowances. Generally, the number of allowances required would be proportionate to the **embedded carbon content** of the import.

Supporters' Views

Some policy analysts, industry stakeholders, climate change experts, and government officials argue that trade measures help address competitiveness and environmental effects associated with domestic emissions pricing systems. For example, some supporters contend that trade measures may do the following:

- *Prevent a decline in output by U.S. producers*: Trade measures would help level the playing field by imposing similar costs on imports from countries without comparable carbon mitigation policies. The potential exists for certain U.S. firms to lose business to energy-intensive imports from countries with weaker climate policies. Moreover, firms could further lose international competitiveness due to the indirect costs of purchasing more expensive U.S. energy. As a result, U.S. firms in vulnerable industries could suffer a loss in output, profits, and employment.

- *Prevent carbon leakage*: Trade measures would prevent U.S. energy- intensive industries from shifting production to countries without comparable carbon mitigation policies, resulting in **carbon leakage**, where emission reductions in the United States are replaced to some degree by production and emission growth in less regulated countries. Supporters of trade measures stated that carbon leakage could lead to higher volumes of greenhouse gas emissions worldwide.

- *Create leverage on other countries to reduce emissions:* Trade measures would create incentives for major trading partners to adopt similar climate policies. For example, proponents of trade measures argue that trade measures could potentially provide leverage to U.S. climate negotiators in their efforts to establish a global framework that includes other major emitting nations. Some industry stakeholders with whom we spoke, for example, stated that the potential benefit of a trade measure in increasing pressure on developing countries may trump its adverse economic effects. These stakeholders said that countries' statements of opposition to proposed trade measures provided evidence of their effectiveness in providing leverage.

- *Political acceptability*: If a climate change bill is to pass, it will be necessary that the bill include a provision like a trade measure to address competitiveness effects. Policy experts stated that most climate change proposals with trade measures have garnered support from industries that could be affected by a domestic emissions pricing system.

Opponents' Views

Some policy analysts, climate change experts, and government officials raise concerns that trade measures may motivate retaliatory actions, undermine efforts to secure multilateral consensus, and generate little leverage. For example, some opponents contend that trade measures may do the following:

- *Motivate retaliatory action*: An important implication of U.S. trade measures is how they would be perceived by other countries and impact negotiations aimed toward an international climate agreement. Opponents argued that trade measures could be viewed by other countries as an antagonistic step, and could lead to retaliatory action from other countries.

- *Undermine efforts to secure international consensus*: While there is general agreement that the global scope of the climate change problem will require actions from all major emitting nations, there are differing views on the impact that trade measures may have in securing an international consensus. Foreign government officials with whom we spoke, declined to give formal positions on trade measure proposals given that a climate change bill has not yet passed Congress. But some officials expressed concerns over the unilateral application of a trade measure by the United States and said that competitiveness effects would be better addressed through multilateral discussions. Because multilateral approaches involve commitments to reduce greenhouse gas emissions by more than one country, they also level the playing field. Multilateral approaches include international agreements, such as the Kyoto Protocol, or international action, as when countries independently decide to reduce their emissions of **greenhouse gases**. Multilateral approaches may involve sectoral agreements, leveling the playing field worldwide within an industry. Foreign embassy officials also stated that trade measures are not likely to be effective tool to leverage other countries to take steps to reduce emissions and that a unilateral U.S. trade measure could add to the challenge of developing multilateral consensus on climate change mitigation efforts.

Table 2. U.S. Imports as a Share of Foreign Output (percentage of metric tons)

	Iron and steel	Primary aluminum	Cement	Pulp products	Nitrogenous fertilizers
Brazil	8.3	5.0	1.1	10.8	3.5
China	0.5	0.3	0.9	0.1	0.5
India	0.7	0.0	0.0	n/a	0.0
Mexico	10.6	n/a	5.6	1.1	8.3
Middle East	0.1	6.5	0.2	0.0	22.1

Source: GAO analysis of production data from the U.S. Geological Survey and the Food and Agriculture Organization and trade data from the Department of Commerce.

Notes to Table

Data are for 2007, except for cement and nitrogenous fertilizers with data for 2006.

Iron and steel production figures include those for pig iron, direct-reduced iron, and raw steel.

Cement production figures are for hydraulic cement.

Pulp products production figures represent the total for "pulp for paper" products within FAOSTAT.

Nitrogenous fertilizer production figures represent tonnage by total nutrients within FAOSTAT.

Not included in the iron and steel import figures above were about 38 million liters from Mexico.

- *Generate little leverage on other countries to reduce emissions:* The effectiveness of trade measures in creating leverage on foreign countries to reduce emissions will vary with the share of their output that is imported into the United States compared with the share that is consumed domestically or exported to other countries. In addition, some policy experts we spoke to note that in some cases, the emission costs that foreign firms will have to pay on their imports into the United States may not be sufficient to motivate them to change domestic production processes. For example, in each of the vulnerable industries listed in table 2, less than 1 percent of total Chinese output is imported into the United States, although the share may be higher for individual firms. Instead, most of the Chinese output in these industries is consumed domestically or in countries that may or may not impose a domestic emissions pricing system. Nonetheless, in certain cases, a U.S. trade measure may be effective in providing some degree of leverage. For example, U.S. iron and steel imports represent more than 8 percent of output in Brazil and more than 10 percent of output in Mexico. In addition, U.S. imports of nitrogenous fertilizers from the Middle East are more than 20 percent of production from that region.

Implementation Challenges

Trade measures could present implementation challenges, particularly with obtaining necessary data to measure the carbon content of imports, and to evaluate and assess climate change actions of other countries. Examples of potential implementation challenges include the following:

- *Determining the embedded carbon content in imports*: Generally, imposing a trade measure on an import would require a determination of the embedded carbon content, or the amount of greenhouse gases emitted during the production of the imported good. According to experts, accurately measuring the **embedded carbon**

content for specific items would be challenging. Furthermore, climate policy experts point out that while determining the embedded carbon content in standardized products like steel, primary aluminum, and basic chemicals would be difficult, it would be even more challenging to assess the embedded carbon content of products that rely on these goods for final assembly.

Additionally, climate policy experts noted that, variations in the type of energy used can result in different carbon intensities for goods that appear identical at the border. To accurately assess the embedded carbon content of a product, the administrator of the program would be required to obtain specific plant-level information on the production process of the import from foreign countries.

Accurately measuring the embedded carbon content of an import would require significant data, resulting in several data reliability concerns with obtaining carbon data from international inventories. Experts and industry stakeholders alike reported that obtaining data from foreign producers could be a challenge, and that using data from international inventories could raise several data reliability concerns as the United States would have no way of verifying that the data obtained was accurate. In addition, under the S. 3036 110TH Cong. (2008), importers would have to submit a declaration statement at the border declaring the imported good is accompanied by the required number of allowances. Agency officials we spoke to noted that officials would not know if the information contained in the declaration was accurate.

- *Evaluating and assessing climate change actions*: As previously discussed, a key feature of a trade measure provision is the requirement of determining the comparability of other countries' actions to address climate change. According to policy experts, among some of the cap-and-trade bills introduced, some have not clearly defined or assessed the method for determining how other countries' actions on climate change would be assess by the U.S. government. In order for the implementing agency to assess the comparability of different countries, the criteria for making this determination would have to be defined either in the legislation or implementing regulations.

Design Trade-Offs

In designing a trade measure, there are a range of possible options to consider as illustrated in table 3. The design features of trade measures involve different trade-offs that will have implications for the effectiveness of the overall measure in addressing competitiveness and environmental concerns.

For example, a key design consideration is the timing for when the trade measure would go into effect. Delaying the implementation of the trade measure could result in the measure being a more effective leverage tool because it would provide other countries with an opportunity to implement similar carbon mitigation policies. On the other hand, industry stakeholders with whom we spoke argued that a delay in implementing the trade measure could adversely affect U.S. industries by incurring costs under a domestic emissions pricing system that their foreign competitors would not face. According to some stakeholders, this could result in U.S. industries being at a competitive disadvantage compared to foreign competitors.

The method for determining the comparability of other countries' actions also involves important trade-offs. Given the different approaches being utilized to address climate change, questions have been raised over whether these different actions would be deemed comparable to U.S. efforts. Industry groups have argued that the method for determining comparability should be based on quantitative criteria that are equivalent to criteria U.S. producers must meet. However, providing flexibility on determining comparability allows countries to utilize the approach best suited for reducing emissions in their individual country.

Table 3. Policy Experts Cite Trade-offs Involved in the Design of Trade Measures

Feature	Options	Design trade-offs
Date when the trade measure goes into effect	The trade measure can go into effect simultaneously when a domestic cap-and-trade system goes into effect or at a later date.	Having the trade measure go into effect simultaneously when the domestic cap-and-trade system goes into effect may reduce impacts on U.S. industry competitiveness. Delaying the implementation of the trade measure allows time for international climate negotiations and for countries to take action on reducing their emissions.
Type of products covered	Climate bills in the 110th Congress limited coverage of the trade measure to specified products such as importers of steel, aluminum, and cement; whereas other bills expanded coverage to manufactured goods that met eligibility criteria.	Limiting product coverage may exclude industries vulnerable to competitiveness effects Expanding product coverage increases federal challenges to track emissions
Defining comparable action	Proposals have varied in stringency and level of detail in outlining the methodology for determining comparable action.	Quantitative criteria for assessing comparable action may not measure the level of effort on climate policy that a country actually undertakes. Flexibility and discretion on comparability determinations increases uncertainty for industry stakeholders.
Calculating border allowance requirements	Proposals have varied in the level of detail in outlining the methodology for calculating the number of allowances industries will be required to submit at the border. Earlier proposals based border allowance requirements on averages for country industry sectors; H.R. 2454 delegated the decision to the future implementing agency.	Basing border allowance requirements on industry average increases implementation feasibility as firm-level data likely not available. Using industry average for baseline may limit incentives to improve efficiency and would penalize efficient firms.

Source: GAO analysis of selected climate change legislation introduced between April 2007 and June 2009 that included trade measures.

3.2. Alternative: Output-Based Rebates

Output-based rebates have been proposed as policy alternatives to using trade measures. These rebates would be provided to energy-intensive (or greenhouse gas-intensive) and trade-exposed industries to cover their increased costs from the carbon pricing system. Under legislative proposals, EPA would determine the average energy (or greenhouse gas) intensity per unit of production for each relevant industrial sector and then distribute allowances based on each facility's amount of production. Although some industry stakeholders note that output-based rebates address competitiveness effects by compensating covered firms for incurred costs, some climate policy experts note that output-based rebates could distort pricing, could reduce incentives for firms to engage in conservation, and may drive up the cost of the program.

Output-based Rebates

Output-based rebates are measures designed to financially rebate industries for the costs incurred under a domestic emissions pricing system. According to policy experts, key features of an output-based rebate include the following:

- *Tied to firm's level of output*: Generally, output-based rebates would be tied to a firm's level of output. For example, firms that expand their operations will receive a larger rebate, while firms that downsize but continue to produce in the United States will receive a smaller rebate. Firms that move offshore or shut down would receive no rebate. A rebate can take the form of a per-unit rebate or tax credit, or a per-unit allocation of emissions allowances.

- *Compensates for incurred costs*: For a facility with an average level of energy intensity (or greenhouse gas intensity) these rebates would cover both direct and indirect costs. These would include their own emissions and costs associated with higher energy prices and technology purchases to improve energy efficiency. Providing allowances based on average industry intensity, energy intensity and production levels, rather than providing them based on each firm's historical emissions, is intended to reward the most efficient facilities and create an incentive for continued efficiency improvements.

H.R. 2454 111th Cong. (2009) included provisions for output-based rebates to offset the costs incurred by firms to comply with a cap-and-trade system. Two other climate change bills, H.R. 7146, 110th Cong. (2008) and H.R. 1759, 111th Cong. (2009), also incorporated output-based rebates. In all three proposals, allowances would be distributed among certain energy- and trade- intensive industries according to an individual firm's output.

Supporters' Views

Some policy analysts, climate change experts, and government officials argue that output-based rebates may address some of the limitations associated with using trade measures. For example, some supporters contend that output- based rebates may do the following:

- *Mitigate costs incurred by U.S. producers*: Output-based rebates are necessary to compensate U.S. firms for costs incurred while complying with a domestic emissions pricing system. Supporters note that trade measures only apply costs to foreign competitors at the border, while output-based rebates affect producers' costs for both domestic and export markets. Among the key supporters of output-based rebates are stakeholders in energy-intensive industries who note that energy costs are a substantial portion of their manufacturing costs. According to these stakeholders, proposals that mitigate costs at the production level, either by allocating **free allowances** to qualifying facilities or rebating the costs of allowances offsets the disincentive created by higher energy prices in the United States.

- *Reward efficient firms*: Output-based rebates are generally allocated in proportion to current levels of production rather than being fixed to historical measures. Supporters argue that because the rebates or allocations are tied to production levels, individual firms have an incentive to increase their production. According to supporters, rebates reward the most productive plants and stimulate investments in efficient technology. Moreover, the rebates would benefit firms that face competition from foreign suppliers in markets at home or global export markets or both.

- *Use U.S. data*: Implementing output-based rebates may be more manageable than implementing trade measures. For example, in proposed legislation, agencies would only require data from U.S. data sources to identify firms that would be eligible for output-based rebates. This differs from trade measures which would require data from foreign firms' production, which could be more difficult to obtain. In addition, legislative proposals have generally been explicit about the U.S. data sources and criteria to be used in identifying eligible industries.

Opponents' Views

Some policy analysts, climate change experts, and government officials argue that despite several advantages identified with using output-based rebates, several potential limitations also exist. For example, some opponents contend that output-based rebates offset the costs of climate policy for sectors that are exposed to unregulated competition, and thereby lower the incentives for those firms to engage in conservation and reduce their energy intensity. Similarly, output-based rebates would likely result in relatively lower output prices for the products produced by firms receiving rebates, thereby reducing the incentives for consumers of those products to conserve. Moreover, some climate policy experts argue that output-based rebates create a subsidy for certain energy-intensive industries which would require non energy-intensive firms to engage in greater efforts to reduce the emissions intensity of their production.

Potential Implementation Challenges

Although the data requirements for an output-based measure may be more manageable than those of a trade measure, some challenges related to identifying electricity distribution data exist. For example, under HR 2454, 111th Cong. (2009), EPA would be required to obtain production data on various commodities from energy-intensive industries, including electricity distribution data. Data on electricity use on a facility-by-facility basis is not currently collected by government agencies. Provisions in either the legislation or in implementing regulations would have to be explicit on how to obtain data on electricity use.

Additionally, according to EPA officials, the methodology outlined in HR 2454, 111th Cong. (2009), for how to determine eligible industries does not align with data that would be collected under EPA's proposed mandatory greenhouse gas reporting rule. According to climate policy experts, standard mechanisms for reviewing eligibility for the rebates are needed to ensure that the policy does not over compensate certain industries.

3.3. Legislative History of Trade Measures and Output-Based Rebates

Between April 2007 and June 2009, several cap-and-trade and carbon tax bills were introduced with provisions to address the potential competitiveness and environmental effects of implementing a domestic emissions pricing system. For example, of the cap-and-trade bills that we examined and that were introduced during that period, some included trade measures such as an allowance requirement at the border or a border tax.

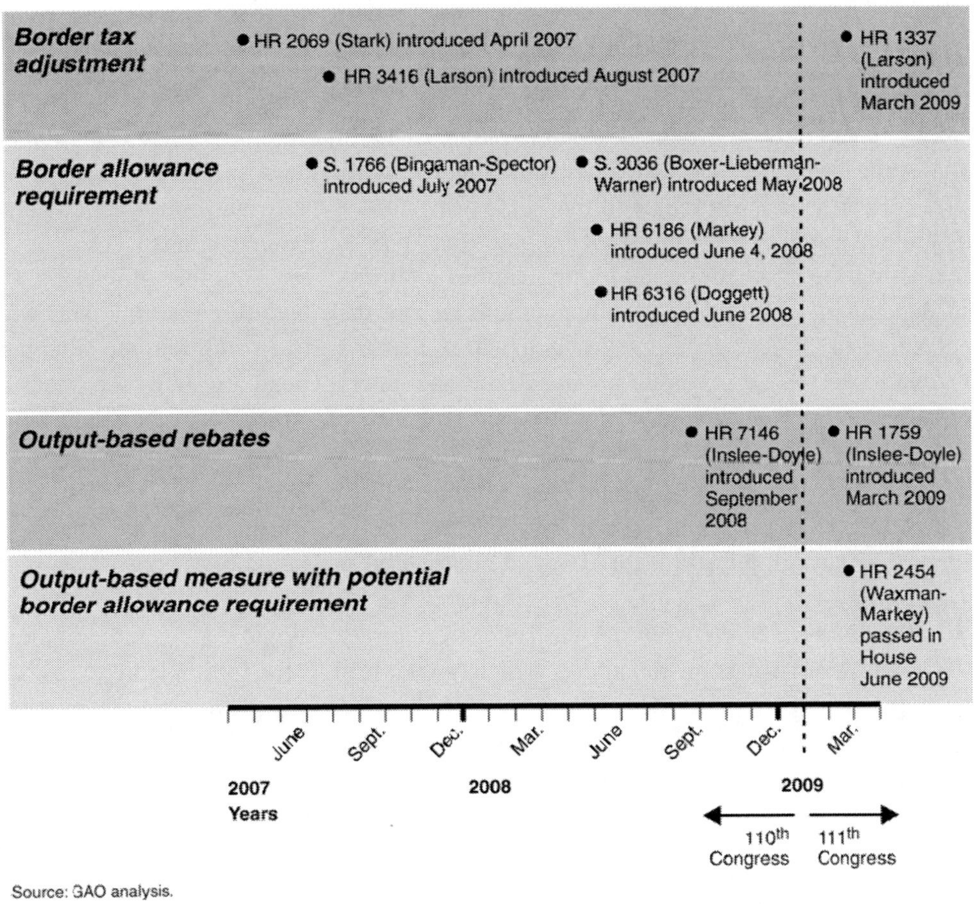

Figure 19. Proposals Move toward Using Output-Based Rebates to Address Competitiveness Concerns

As illustrated in figure 19, two bills introduced in 2009 have included provisions for using output-based rebates. The most recent bill, H.R. 2454, 111th Cong. (2009) includes a provision for output-based rebates. Under that bill, energy intensive and trade exposed industries would receive allowances to cover their increased costs.

Several industry stakeholders with whom we spoke, reported that output-based rebates and trade measures could function as complimentary measures to address competitiveness concerns. For example, in H.R. 2454, 111th Cong. (2009), the output-based rebates would be phased out by 2035 unless the President decides to extend them. In that bill, the trade measure would go into effect after 2020, unless the President determines it would not be in the national interest and Congress passes an affirmative joint resolution within 90 days.

Some stakeholders have stated that having the trade measure to go into effect at a later time may provide time for the United States to be part of international negotiations on climate change.

4. International Trade Implications: Overview

Introduction

In this section, we discuss potential international implications of a U.S. trade measure, such as concerns about countries possibly retaliating in response to a trade measure or its potential impact on the multilateral trading system. In addition, we provide information on the WTO dispute settlement process, a forum that facilitates the resolution of specific trade disputes. We provide information on the WTO provisions and key questions that may apply if a U.S. trade measure or an output-based rebate were to be challenged at the WTO.

Assessing the international trade implications is difficult for a number of reasons. One is that it depends in part upon how other nations reduce their carbon emissions, and whether they perceive any U.S. measures as likely to affect their exports. In addition, the outcome of a WTO challenge, if any, would be uncertain and may depend on how the measure is implemented.

Potential Bilateral and Multilateral Trade Impacts

According to experts and foreign officials we spoke with, U.S. climate change trade measures may have implications for U.S. and multilateral trade relations.

- *Potential for bilateral trade retaliation:* Countries may view U.S. trade measures as trade restrictions or sanctions, which could lead them to implement restrictions against U.S. exports. Other countries could potentially develop counter measures based on a different test of "comparability," such as historical or per capita emissions, according to trade experts we spoke with.

  Although no other country has yet to implement a trade measure based on greenhouse gas emissions, European officials have previously considered the use of trade measures on countries, including the United States, that have not taken actions that match European standards. In addition, support within other countries for imposing trade measures could increase if the United States implements a trade measure.

- *Potential for increased trade tensions:* A broader concern is that escalating retaliation between the United States and other countries could lead to significant global trade tensions. Given the volume of trade in goods that could potentially be affected by trade measures linked to greenhouse gas emissions, some officials and experts have argued that escalating tensions and responses to these measures could ultimately do significant harm to the functioning of the multilateral trading system under the WTO. Trade experts told us that trade measures and responses could potentially pose systemic challenges to the WTO and its dispute resolution system. For example, multiple challenges brought up by other countries based on different imported products could tie up the system.

4.1. WTO Compliance Questions

WTO Dispute Settlement Process

The WTO's dispute settlement system facilitates the resolution of specific trade disputes and serves as a vehicle for upholding trade rules and preserving the rights and obligations of WTO Members under WTO agreements. WTO Members may request consultations concerning measures affecting the operation of such WTO agreements. After consultations, WTO Members can request the establishment of a panel to hear their claims regarding alleged inconsistencies of other Member's measures. The panel will issue a report of its findings, which will be adopted by the WTO **Dispute Settlement Body (DSB)**, unless a party to the dispute appeals the panel's report, which is reviewed by the **WTO Appellate Body**. WTO disputes can take several years or more to complete the entire process. If the responding party does not prevail and fails to comply with the rulings and recommendations of the DSB, the complaining party may seek authority to suspend concessions or other obligations under the WTO agreements.

WTO DISPUTE SETTLEMENT PROCESS

Consultation (up to 60 days)

Panel Review (generally up to 6 months, but in no case more than 9 months)

Appellate Stage (60 to 90 days)

DSB Decision (generally up to 9 months and up to 12 months if the panel report is appealed)

Implementation Stage (if immediate compliance is impractical, member given a 'reasonable period of time' to comply, which normally should not exceed 15 months from adoption of report up to 15 months)

A unilateral U.S. trade measure or system of rebates linked to a climate change policy may be challenged by other countries through the WTO's dispute resolution process. Countries could potentially raise different types of challenges depending on the design of the trade measure or rebate, and several key questions may be relevant in the consideration of a WTO dispute. These questions stem from the WTO provisions under which the trade measures or rebates may be challenged and reviewed, including the General Agreement on Tariffs and Trade (GATT), and the Agreement on Subsidies and Countervailing Duties (ASCM).

The following pages describe in more detail the questions, factors, and WTO provisions listed in table 4.

The following observations are based on comments from various legal scholars and other non-governmental commentators on questions and provisions that may be relevant in analyzing trade measures contained in climate change legislation for consistency with WTO rules.

1. **Is the trade measure consistent with WTO market access rules and commitments?**

   An initial question involves how a trade measure would be interpreted and the corresponding GATT article under which the measure would be reviewed.

   - GATT Article II:1(b) prohibits a WTO Member from applying customs duties or other duties or charges to imports in excess of the Member's schedule of tariff concessions.
   - GATT Article XI: 1 prohibits prohibitions or restrictions, other than duties, taxes or other charges, on imports, with certain exceptions.
   - According to legal experts, a "border tax" imposed on imports equivalent to a tax imposed on like domestic products may be WTO compliant.

2. **Is the trade measure consistent with WTO non-discrimination requirements?**

   Trade measures must also be consistent with WTO non-discrimination provisions covering most favored nation status treatment and national treatment.

   - *Does a trade measure treat two like products differently depending on country of origin?* GATT Article I:1 provides, in part, that with respect to customs duties and charges imposed on or in connection with importation and exportation, and the method of levying such duties and charges, any advantage granted by any WTO Member to any product originating in any other Member is to be accorded immediately and unconditionally to the like product originating in or destined for all other WTO Members.

   - *Does a trade measure treat imported products less favorably than like domestic products?* GATT Article III:4 provides, in part, that imported products shall be accorded treatment no less favorable than that accorded to like products of national origin in respect of all laws, regulations, and requirements affecting their internal sale, offering for sale, purchase, transportation, distribution, or use.

3. **If the trade measure is not permissible under core WTO obligations, would it be covered under an environmental exception?**

   If a trade measure does not comply with GATT market access or nondiscrimination provisions, it may still be justified under certain GATT environmental exceptions. A key consideration would be the interpretation of the intended purpose of the measure and whether it is designed to serve an environmental objective.

- *Is the trade measure designed to serve an environmental objective?*
  - GATT Article XX (b) provides an exception for measures necessary to protect human, animal, or plant life or health.
  - GATT Article XX (g) provides an exception for measures relating to the conservation of exhaustible natural resources if such measures are made effective in conjunction with restrictions on domestic production or consumption.

- *How is the measure applied?* The introduction (chapeau) to GATT Article XX states that such measures must not be applied in a manner that would constitute a means of arbitrary or unjustifiable discrimination or a disguised restriction on international trade.

4. **Are output-based rebates consistent with WTO rules governing subsidies?** The ASCM imposes disciplines on certain kinds of subsidies to domestic producers, such as those that cause adverse effects to foreign competitors.

- *Is an output-based rebate or distribution of free allowances a subsidy?* To be considered a subsidy under the ASCM, a system of rebates would need to involve (1) a "financial contribution" by the government or any form of income or price support that (2) confers a "benefit" to the recipient and (3) is "specific."

- *Is the rebate an "actionable" subsidy that causes "adverse effects" to the interests of other WTO Members?* A further key question is whether a rebate, if found to be a subsidy, would cause "serious prejudice" to the interests of other WTO Members.

Shrimp-Turtle Case- In this 1996 case, four WTO members challenged as discriminatory a U.S. embargo on the importation of certain shrimp and shrimp products from countries that had not been certified by the Department of State as having a comprehensive sea turtle conservation regime. The United States argued that the measure was justified under article XX (g).

However, the WTO panel found, in part, that the U.S. measure was not justified under article XX (g) because it met the panel's test of being a "risk to the multilateral trading system." Upon appeal, the Appellate Body rejected the panel's test and found that the measure fell within article XX (g). However, the Appellate Body also found the U.S. measure had been applied in an arbitrary and discriminatory manner.

The United States was ultimately able to comply with WTO rulings by making changes in the way the law was administered, including revising guidelines to establish greater transparency and due process in country certification decisions.

Table 4. Key Questions That May be Considered in a WTO Challenge

Key questions	Discussion	WTO provisions
1. Is the trade measure consistent with WTO market access requirements?	Customs duties or other duties or charges on imports in excess of schedules of concessions may not be imposed. Subject to exceptions, prohibitions or restrictions, other than duties, taxes or other charges, on imports may not be imposed.	GATT Article II, XI
2. Is the trade measure consistent with WTO non- discrimination requirements?	Measure must treat imported products no less favorably than like domestic products.	GATT Article III
	Measure must also not discriminate among goods from different countries.	GATT Article I
3. If not WTO consistent, is the trade measure covered by a WTO environmental exception?	Measure relating to conservation of exhaustible natural resources or that is necessary to protect human, animal or plant health is covered under WTO exceptions.	GATT Article XX (b), (g)
	To be covered, measure must not be applied in a manner which would constitute a means of arbitrary or unjustifiable discrimination or be a disguised restriction on international trade	GATT Article XX chapeau
4. Are output-based rebates consistent with WTO rules governing subsidies?	The rebates must not involve a specific subsidy or other benefit that causes adverse effects to other WTO Members.	ASCM

Source: GAO analysis of legal scholars' views on questions and provisions that may be relevant in analyzing trade measures contained in climate change legislation for consistency with WTO rules.

APPENDIX 1. OBJECTIVES, SCOPE AND METHODOLOGY

Our research objectives were to examine: (1) information on estimating industry effects; (2) examples of industries that may be vulnerable to a loss in international competitiveness from **emissions pricing**; (3) trade measures and other approaches to address competitiveness issues; and (4) potential international implications of trade measures.

To address all of these objectives, we (1) reviewed relevant climate change academic literature and federal agency documents; (2) reviewed and analyzed studies and data on climate change from a variety of sources, including past GAO work, the U.S. Census Bureau and Departments of Commerce and Energy, international organizations, policy institutes, and universities; (3) interviewed agency officials from the Office of the U.S. Trade Representative (USTR); the Environmental Protection Agency (EPA); the Departments of Commerce, State, Energy, and Treasury, and the Department of Homeland Security's Customs and Border Protection (CBP); (4) conducted interviews with subject-matter experts selected from a population of individuals from government, academia, business, and professional organizations. Several criteria were used for selecting experts to interview including: type and depth of experience, the expert's recognition in the professional community, relevance of published work, professional affiliations, present and past positions held, and other subject matter experts' recommendations.

Objective 1 and 2

To better understand how implementing a domestic emissions pricing system would likely affect the international competitiveness of U.S industries, we reviewed relevant studies from GAO, EPA, the Department of Energy's Energy Information Agency (EIA), and international organizations. We also reviewed documents and interviewed experts from policy institutes, universities, and industry.

Recently proposed legislation identifies industries potentially vulnerable to adverse competitiveness effects as those with: (a) either energy intensity or greenhouse gas intensity of 5 percent or greater; and (b) a trade intensity of 15 percent or greater.[4] Since calculation of greenhouse gas intensity requires an emissions price—not yet determined—and detailed industry data on greenhouse gas emissions—not currently available—we focus our analysis on industry energy intensity. However, energy intensity is also an imperfect measure of potential vulnerability since energy use is not the only source of industrial greenhouse gas emissions, and an industry's energy expenditures may rise with to a change in market prices, even if the industry's greenhouse emissions decline. Additionally, many factors may determine the vulnerability of any particular firm, including transportation costs, supply chain relationships, and access to markets for natural resources, capital, or labor.

Based on industry and expert studies, most industries that meet both vulnerability criteria are within primary metals, nonmetallic minerals, paper products, and chemicals.[6] Quantitative analysis of the estimated competitiveness impacts of U.S. greenhouse gas emissions pricing is relatively limited. We present results for two studies that separate the international competitiveness effects from U.S. emissions pricing from the domestic market effects. Estimated results from these two studies include findings that are broadly consistent with those of other studies. For example, a study of climate policy impacts on U.S. industries was commissioned by the National Committee on Energy Policy and studies of the European Union's climate policies have been performed for the European Commission and the International Energy Agency. In each of these studies, the list of vulnerable industries is generally consistent, as is the finding that estimated impacts could be greater for energy-intensive industries.

To illustrate variation in economic characteristics among these four manufacturing industries, we analyzed: (1) energy consumption data from the U.S. Census Bureau's Annual Survey of Manufacturers; (2) energy use data from EIA's Manufacturing Energy Consumption Survey; (3) international trade data from the Department of Commerce's International Trade Administration; and (4) U.S. and international production data from the Department of Commerce's Bureau of Economic Analysis, the U.S. Geological Survey and the United Nations Food and Agricultural Organization (FAO). For analysis of sub-industries, we examined data with a Census Bureau North American Industry Classification System (NAICS) code of 5 to 6 digits, depending on how the data were reported in each of the databases reviewed and for consistency of comparison with other studies. For the vulnerable sub- industry charts, we applied multiple criteria to show variation in industry characteristics of energy intensity, trade intensity, and primary trading partners. For example, we selected sub-industries that met both the energy and trade intensity criteria, examples that met only one criterion, and examples that met neither, but had significant imports from **non-Annex I** countries. We do not identify the complete list of potentially vulnerable subindustries, a list that will vary depending on the level of aggregation in product lines that is examined. The specific list of example sub-industries we selected, along with their corresponding NAICS codes can be found in table 5.

We tested the data we analyzed for internal consistency and for consistency with other key published studies, interviewed agency and industry officials regarding appropriate use of the data, and reviewed source data information regarding the entity's methodology and actions taken to ensure data reliability. We determined that the data were sufficiently reliable for our use. We note, however, that industry characteristics may change over time and may vary for firms within each sub-industry selected. To supplement our data analysis, we conducted interviews with industry groups such as steel, aluminum, chemicals, and the Energy-Intensive Manufacturer's Working Group.

Objective 3

To identify and compare key features of proposed trade measures, we reviewed and analyzed selected climate change legislation introduced between 2007 and 2009, and congressional hearing records. From these documents we extracted information regarding features of the trade measures, such as objectives, scope of coverage, effective date for trade measures, and other key features. To identify how features of trade measures could address the potential economic and environmental effects of emissions pricing, we reviewed studies and reports obtained from our literature review, including past GAO reports on climate change and studies published by climate change policy institutes such as Resources for the Future, the Peterson Institute, and the Pew Center on Global Climate change. We also interviewed subject matter experts and agency officials from USTR; EPA; the Departments of Commerce, State, Energy, and Treasury, and CBP, regarding these issues and to identify the potential implementation and administrative challenges with using trade measures. In our interviews with agency officials we discussed steps agencies have taken to anticipate implementation challenges. To obtain information about climate change policies in other countries and to learn about the potential impact of U.S. trade measures on bilateral relations and international negotiations, we also interviewed officials from the embassies of Australia, Brazil, Canada, China, European Union, Mexico, and Singapore.

Objective 4

To assess potential international implications of using trade measures, we interviewed subject matter experts, agency officials in EPA, CBP, USTR, EIA, and the Departments of Treasury, State, and Commerce, and foreign embassy officials. To discuss questions and provisions that may be relevant in analyzing trade measures and output-based rebates for consistency with WTO rules, we reviewed WTO documents, obtained information from our interviews with subject-matter experts and agency officials, and reviewed information obtained from our literature review.

Table 5. Sub-Industries Selected as Vulnerable Industry Examples

Sub-Industry Examples	North American Industry Classification System Codes
Primary Metals	**331**
Iron and steel mills	331111
Electrometallurgical products	331112
Steel manufacturing	33121, 33122
Ferrous metal foundries	33151
Primary aluminum	331312
Aluminum products	331314, 331316
Nonmetallic Minerals	**327**
Cement	32731
Concrete	32732
Lime and gypsum	3274
Glass	32721
Clay building material	32712
Mineral wool	327993
Paper Products	**322**
Pulp mills	32211
Paper mills	32212
Paperboard mills	32213
Paperboard containers	32221
Chemicals	**325**
Alkalies and chlorine	325181
Carbon black	325182
Nitrogenous fertilizers	325311
Industrial gases	32512
Artificial and synthetic fibers	32522

Source: U.S. Census Bureau, http://www.census.gov/eos/www/naics/

USTR provided technical comments on this report.

We conducted our work from October 2008 to July 2009 in accordance with all sections of GAO's Quality Assurance Framework that are relevant to our objectives. The framework requires that we plan and perform the engagement to obtain sufficient and appropriate evidence to meet our stated objectives and to discuss any limitations in our work. We believe that the information and data obtained, and the analysis conducted, provide a reasonable basis for any findings and conclusions in this product.

APPENDIX 2. NOTES TO INDUSTRY FIGURES

Notes to Figure 2: Energy intensity is calculated as the value of purchased fuels and electricity as a share of the value of output. Trade intensity is calculated as the value of total trade (exports plus imports) divided by the value of output plus imports. Data for value of output are from the Bureau of Economic Analysis Gross Domestic Product (GDP) by industry accounts, shipments by industry in current dollars.

Notes to Figure 3: Due to aggregation in source data, energy intensity for electrometallurgical products is assumed to be the same as iron and steel mills. Energy intensity data for primary aluminum is based on 2005 data. Aluminum products include industries characterized by the North American Industry Classification System (NAICS) with codes 331314 and 331316 and steel manufacturing industries represent codes 33121 and 33122.

Notes to Figure 4: Energy types characterized as "Renewables and other" includes residual and distillate fuel oil, liquefied petroleum gases, natural gas liquids, net steam (the sum of purchases, generation from renewables, and net transfers), and other energy that respondents indicated was used to produce heat and power or as feedstock/raw material inputs. Net electricity is the sum of purchases, transfers in, and generation from noncombustible renewable resources minus quantities sold and transferred out. Due to aggregation in source data, "Aluminum" includes primary aluminum and aluminum products.

Notes to Figure 5: "EU Plus" includes countries from the EU and other Annex I countries not represented elsewhere. Other than China, the largest two sources of imports for electrometallurgical products are the Republic of South Africa (17 percent) and Trinidad and Tobago (10 percent).

Notes to Figure 6: The current value of imports from all countries is shown in the table below the chart.

Notes to Figure 7: Energy intensity data for clay building materials are based on 2005 data because 2006 data were not available. Glass products include industries characterized by the North American Industry Classification System (NAICS) with codes 32721.

Notes to Figure 8: Energy use by type data for lime products excludes gypsum. Data on energy use by nonmetallic mineral sub-industry examples excluded from the chart are not currently reported by the Department of Energy.

Notes to Figure 9: After Canada and China, the two largest sources of imports for cement are Korea (9 percent) and Colombia (8 percent).

Notes to Figure 10: The current value of imports from all countries is shown in the table below the chart.

Notes to Figure 11: Due to aggregation in Bureau of Economic Analysis value of output data for paper mills with paperboard mills, value of output data for these two sub-industries are for 2006 and are from the Census Bureau's Annual Survey of Manufacturers, shipments by industry in current dollars. Aggregated data for 2006 are consistent between the two sources.

Notes to Figure 12: Data for coal and coke usage by pulp mills are not publicly available to avoid disclosure of individual firm information, such that percentage shares could not be computed. Data on energy use by paperboard containers is not currently reported by the Department of Energy.

Notes to Figure 14: The current value of imports from all countries is shown in the table below the chart.

Notes to Figure 15: Energy intensity data for alkalies and chlorine and carbon black is based on 2005 data because 2006 data is currently not available.

Notes to Figure 16: Data for certain energy uses by type for alkalies and chlorine, industrial gases, and artificial and synthetic fibers is not publicly available to avoid disclosure of individual firm information, such that percentage shares could not be computed.

Notes to Figure 17: Other than Canada, the two largest sources of imports for nitrogenous fertilizers are Trinidad and Tobago (26 percent) and the Middle East (14 percent), with countries in the Middle East defined by the U.S. Department of State.

Notes to Figure 18: The current value of imports from all countries is shown in the table below the chart.

Notes to Figure 19: In addition to these bills, Representatives John Dingell and Rick Boucher released a discussion draft on October 7, 2008. The discussion draft included a cap-and-trade proposal, and presents four options for allocating allowances under a cap-and-trade system.

Appendix 3. Works Cited

Aldy, Joseph E. and Pizer, William A. (May, 2009) "*The Competitiveness Impacts of Climate Change Policies.*" Pew Center on Global Climate Change, Arlington, VA.

Fischer, Carolyn and Fox, Alan K. (February, 2009) "*Comparing Policies to Combat Emissions Leakage: Border Tax Adjustments versus Rebates.*" RFF Discussion Paper No. 09-02, Resources for the Future, Washington, D.C.

GAO, *Climate Change: Expert Opinion on the Economics of Policy Options to Address Climate Change*, GAO-08-605 (Washington, D.C.: May 9, 2008).

GAO, *International Climate Change Programs: Lessons Learned from the European Union's Emissions Trading Scheme and the Kyoto Protocol's Clean Development Mechanism*, GAO-09-151 (Washington, D.C.: November 18, 2008).

Ho, Mun S., Richard Morgenstern, and Jhih-Shyang Shih. (November, 2008) *"Impact of Carbon Price Policies on U.S. Industry."* RFF Discussion Paper No. 08-37, Resources for the Future, Washington, D.C.

U.S. Environmental Protection Agency, Office of Atmospheric Programs. *EPA Preliminary Analysis of the Waxman-Markey Discussion Draft,* The American Clean Energy and Security Act of 2009 in the 111th Congress, Apendix. April 20, 2009.

Appendix 4. Climate Change Bills Cited In Report

Save Our Climate Act of 2007, H.R. 2069, 110th Cong., (2007)

America's Energy Security Trust Fund Act of 2007, H.R. 3416, 110th Cong., (2007)

Low Carbon Economy Act of 2007, S. 1766, 110TH Cong., (2007)

Lieberman-Warner Climate Security Act of 2008, S. 3036, 1 10TH Cong., (2008)

Investing in Climate Action and Protection Act , H.R. 6186, 110TH Cong., (2008)

Climate Market, Auction, Trust & Trade Emissions Reduction System Act of 2008, H.R. 6316, 110TH Cong., (2008)

Carbon Leakage Prevention Act , H.R. 7146, 110TH Cong., (2008)

America's Energy Security Trust Fund Act of 2009, 11 1TH Cong., (2009)

Emission Migration Prevention with Long-term Output Yields Act , H.R. 1759, 111TH Cong., (2009)

American Clean Energy and Security Act of 2009, H.R. 2454, 111TH Cong., (2009)

Appendix 5. Glossary

- **Annex I Countries:** Parties to the United Nations Framework Convention on Climate Change (UNFCCC) that are industrialized countries and were members of the Organization for Economic Cooperation and Development (OECD) in 1992 plus countries characterized as economies in transition.

- **Allowances**: In the context of a cap-and-trade system, an emissions allowance is a permit to emit a specific quantity of emissions.

- **Border allowance requirement**: A measure that would require importers to purchase allowances from the federal government prior to importing goods into the United States. **Border tax adjustment**: A levy on imported goods proportionate to the imports' embedded carbon content. Generally, the levy on imported goods would be equivalent to the tax applied to domestic goods under a carbon tax system. Three legislative bills-- H.R. 2069, 110TH Cong. (2007), H.R. 3416, 110TH Cong. (2007), and H.R. 1337-- 111TH Cong. (2009), include provisions for a tax on any taxable carbon substance sold by the importer. For the purposes of this report, we are using the term border tax adjustment to refer to this tax described in the bills and to differentiate it from a carbon tax applied to domestic producers.

- **Business-as-usual**: A scenario in which no action is taken to reduce greenhouse gas emissions.

- **Cap-and-trade**: An emissions pricing system in which the government applies an aggregate cap or quota to limit total emissions from regulated sources, which are required to hold allowances to cover their emissions. Allowances are allocated by the government and may be traded. Sources whose allowances exceed their emissions may offer permits for sale, while sources for which emissions exceed allowances will need to buy them.

- **Carbon dioxide equivalent**: The quantity of carbon dioxide emissions that would trap as much heat as a quantity of non-carbon-dioxide gases, such as methane, sulfur hexafluoride, nitrous oxide, or industrial gases.

- **Carbon leakage**: The condition when emissions reductions in one country are replaced by increases in emissions in other countries.

- **Carbon offsets**: Reductions in greenhouse gas emissions from an activity in one place to compensate for emissions released elsewhere.

- **Carbon tax system:** A system that requires regulated sources to pay a charge based on the level of their emissions.
- **Competitiveness:** The ability of a U.S. industry to compete successfully in international markets with foreign competitors.

- **Cost containment measures**: Mechanisms designed to reduce the economic impact of climate change legislation on certain regulated entities and provide them with flexibility in managing compliance costs. Examples include banking and borrowing of allowances, price caps on allowances, free allocation of allowances, and carbon offsets.

- **Dispute Settlement Body**: The WTO's dispute settlement process is administered by the Dispute Settlement Body, composed of representatives from WTO Members, and rules set time limits for each step in the process.

- **Downstream regulation**: Regulation on sources that emit greenhouse gases.

- **Embedded carbon content**: Carbon emissions associated with theproduction of a product through the entirety of its supply chain.

- **Emissions pricing**: A market-based mechanism, such as a carbon tax or cap-and-trade system, to encourage reductions in emissions by putting a price on them. In this report, "emissions pricing" refers to greenhouse gas emissions pricing, as opposed to pricing systems for other types of emissions.

- **Energy intensity:** The industry's cost of purchased electricity and fuel costs, or energy expenditures, divided by the value of shipments (output) of the industry, as defined in H.R. 2454.

- **Free allowance allocation:** Emission allowances given by the government for free. Under a cap-and-trade program governments can either give allowances to regulated entities for free or they can sell allowances through an auction. Allocating allowances for free represents a transfer of wealth from the government to the entities receiving the allowances, while auctioning allowances enables the government to decide how to use the revenue.

- **Greenhouse gas intensity:** Twenty times the quantity of carbon dioxide equivalent emissions from a sector, divided by the value of shipments (output) for the sector, as defined in H.R. 2454.

- **Greenhouse gases:** Gases such as carbon dioxide, methane, nitrous oxide, and other substances that increase temperatures by trapping heat that would otherwise escape the earth's atmosphere.

- **Intergovernmental Panel on Climate Change (IPCC):** A scientific intergovernmental body set up by the World Meteorological Organization (WMO) and by the United Nations Environment Program (UNEP). The IPCC was established to provide decision makers and others interested in climate change with an objective source of information about climate change.

- **International reserve allowance:** Under a cap-and-trade system, the United States could require importers to acquire emissions allowances corresponding to the level of greenhouse gases emitted during production.

- **Output-based rebates**: Rebates based on actual, recent measures of production that are given to regulated sources in a cap-and-trade system for allowances they have purchased to cover their emissions.

- **Trade intensity:** The ratio of the sum of the value of imports and exports within an industry to the sum of the value of shipments (output) and imports within an industry, as defined in H.R. 2454.

- **Trade measures**: Cost equalization measures at the border that impose a cost or other requirement on energy-intensive imports from countries with weaker climate policies. Depending on the type of domestic carbon mitigation system in place, trade measures can take several forms. For example, trade measures can be proposed as part of a cap-and-trade system (allowance requirement at the border) or a carbon tax system (border tax).

- **Upstream regulation**: Greenhouse gas regulation focused on the sale of fuels that produce greenhouse gases when they are used.

- **WTO Appellate Body**: A body within the WTO's that reviews a WTO panel's legal findings during a dispute resolution process. The Appellate Body and the Appellate Body's report is to be accepted by parties in the dispute unless the Dispute Settlement Body decides by consensus not to adopt the report.

GAO's Mission

The Government Accountability Office, the audit, evaluation, and investigative arm of Congress, exists to support Congress in meeting its constitutional responsibilities and to help improve the performance and accountability of the federal government for the American people. GAO examines the use of public funds; evaluates federal programs and policies; and provides analyses, recommendations, and other assistance to help Congress make informed oversight, policy, and funding decisions. GAO's commitment to good government is reflected in its core values of accountability, integrity, and reliability.

End Notes

[1] A cap-and-trade system is an emissions pricing system in which the government applies an aggregate cap or quota to limit total emissions from regulated sources, which are required to hold allowances to cover their emissions. Allowances are allocated by the government and may be traded. Sources whose allowances exceed their emissions may offer permits for sale, while sources for which emissions exceed allowances will need to buy them.

[2] Under a carbon tax system, regulated sources pay a charge based on the level of their emissions.

[3] You also asked GAO to examine revenue measures under consideration as part of climate change legislation. GAO plans to issue this product later in 2009.

[4] For example, H.R. 2454 (111th Cong.), and H.R. 1759 (111th Cong.) include these criteria. HR 2454 also identifies industries with an energy intensity of 20 percent or greater as vulnerable to adverse competitiveness effects irrespective of trade intensity; however, none of the examples we discuss in our analysis would be identified as vulnerable under these criteria.

[5] A study commissioned by energy-intensive industries identifies 41 manufacturing sub-industries as meeting these vulnerability criteria in at least one year between 2004 and 2006, of which 34 subindustries are within primary metals, nonmetallic minerals, paper products, and chemicals. We did not conduct sensitivity analysis to determine how many additional sub-industries would be identified as vulnerable if the criteria were defined

more broadly. According to another expert, the list of potentially vulnerable sub-industries is greater if using the proposed energy-intensive criteria compared with carbon-intensity criteria using a per ton carbon price of $20 or $30. Further, eliminating the trade-intensity criteria would mean that roughly an additional 10 sub-industries would be identified as vulnerable, whereas eliminating the energy-intensity criteria would yield over 150 subindustries that would be identified as vulnerable to competitiveness effects.

INDEX

D

E

N

O

P

Q

R

S

T